72nd Tobacco Science Research Conference (TSRC 2018)

Memphis, Tennessee, USA
16-19 September 2018

ISBN: 978-1-5108-8889-0

Printed from e-media with permission by:

Curran Associates, Inc.
57 Morehouse Lane
Red Hook, NY 12571

Some format issues inherent in the e-media version may also appear in this print version.

Printed by Curran Associates, Inc. (2019)

TSRC Conference Committee
http://www.tsrcinfo.com
info@tsrcinfo.com

Additional copies of this publication are available from:

Curran Associates, Inc.
57 Morehouse Lane
Red Hook, NY 12571 USA
Phone: 845-758-0400
Fax: 845-758-2633
Email: curran@proceedings.com
Web: www.proceedings.com

TABLE OF CONTENTS

Importance of science to FDA's Center for Tobacco Products

HOLMAN M.R.
U.S. Food & Drug Administration, Silver Spring, MD, USA

As regulators of tobacco products, the U.S. FDA's Center for Tobacco Products (CTP) is a science-based agency. Since its inception, CTP's Office of Science (OS) has worked to build scientific knowledge around tobacco products in order to meet its public health mission. OS has not only increased the scientific knowledge of its staff but also puts considerable resources toward ensuring that all stakeholders have access to that knowledge. For example, OS staff co-author numerous publications in scientific journals each year. In addition, CTP funds research that is presented at scientific conferences and published in scientific journals regularly. CTP uses this ever-increasing scientific knowledge of tobacco products to make well-informed regulatory decisions founded in the best available science. The scientific basis for CTP's decisions is evident in our actions (e.g., application review, rulemaking, guidance, and enforcement actions).

A risk-based approach to tobacco product development & control

CIAMBRONE K.
ITG Brands LLC, Greensboro, NC, USA

The Tobacco Industry does not appear to have a model that identifies those attributes of a product's characteristics and manufacturing processes having the greatest impact on a finished tobacco product's quality and compliance. As a result, regulators seem to consider all attributes of a tobacco product with the same criticality. This presentation proposes that a flexible regulatory framework for tobacco products could be achieved by utilizing a risk-based end-to-end control strategy similar to that established in the pharmaceutical industry. A control strategy is the description of the product, manufacturing process, facilities and equipment, and assurance elements that provides confidence in the consistency and quality of a product. The first step of this approach is to clearly define the adult consumer's expectations of quality for a product's intended use. Only after that is defined, can a risk-based approach to tobacco product development and delivery enable a fit-for-purpose control strategy commensurate to expected quality, complexity and required compliance. An effective control strategy identifies product characteristics that are critical to meeting these predefined expectations. These critical characteristics are then evaluated by assessing the extent to which their variation can have negative impact on the finished tobacco product. Proportional controls are then designed in order to achieve the quality as defined in the first step. Once controls are identified and established, those critical attributes, or the process controls, should then be monitored to ensure that the product is consistently meeting quality and compliance expectations. Using a risk-based approach ensures that the quality for intended use, product characteristics and manufacturing and supply chain complexity is considered in order to differentiate between simple products that may only require end-product testing and complex products that may require in-process controls and more release testing. It should be noted that a control strategy, for any tobacco product, should reflect of the level of maturity and effectiveness of the quality management system in place. As in the pharmaceutical industry, it should also be recognized that the quality of knowledge gained and provided to a regulatory agency, and not the quantity of data, should be the basis for regulatory evaluation. Ultimately, a well-articulated and executed control strategy can consistently deliver the intended quality and compliance of a tobacco product. This approach should be considered as part of a reasonable regulatory framework that provides value for the adult consumers by ensuring consistent quality while providing options and access to new harm reduction products by reducing regulatory burden and cycle times. This proposal is based on International Standards on Harmonization (ICH) Quality Guidelines (Q8, Q9 & Q10). References: • PHARMACEUTICAL

The role of non-combusted products in tobacco harm reduction

FLORA W.; GOGOVA M.; SARKAR M.A.
Altria Client Services LLC, Richmond, VA, USA

Today, FDA has regulatory authority over all tobacco products. FDA acknowledges that there is a continuum of risk for tobacco products and distinguishes between the harm associated with combustible versus non-combustible products. A harm-reduction strategy that informs adult smokers about reduced-risk products, subject to FDA oversight, will complement, not compete, with proven prevention and cessation strategies. This approach should focus on reducing tobacco-related morbidity and mortality among the population of adults who continue to use tobacco products by empowering them to make an informed decision to choose a product proven to be lower on the continuum of risk. That is why we invest in developing a portfolio of non-combustible tobacco products. We've been concentrating on three product platforms. These include smokeless and other oral nicotine containing products, e-vapor, and heated tobacco products. We believe adult smokers unable or unwilling to quit are more likely to completely switch from cigarettes if they can choose from a variety of acceptable alternative tobacco products. Of course, we can only compete in the marketplace with products authorized by the FDA and we can only communicate reduced harm or risk claims if FDA grants authorization to a Modified Risk Tobacco Product Application (MRTPA). As manufacturers, it is critical to develop a variety of innovative reduced harm tobacco products that are acceptable alternatives to conventional cigarettes for adult smokers, transparently provide extensive science and evidence about these products, and, with FDA authorization, bring them to market. FDA and the broader public health must provide reasonable regulatory pathways, unbiased scientific assessments, collaborative research, and science and evidence based differential risk communications about these products. A diverse market of FDA-authorized, non-combustible products with accompanying modified risk claims will enable informed decisions for adult cigarette consumers to choose products proven to be lower on the continuum of risk and thus is an extremely important step in advancing tobacco harm reduction in the U.S.

The impact of FDA regulation, a CRO's perspective on changes in the industry

GILLMAN I.G.
Enthalpy Analytical LLC, Durham, NC, USA

The signing of the Tobacco Control Act (TCA) on June 22, 2009 caused far reaching and permanent changes to the regulation and testing of tobacco products in the US. These changes have been seen in all aspects of the industry including manufacturing, labeling, product development and also analytical testing. The TCA created a framework for tobacco product regulation however, did not include details on how products would be regulated. The TCA also outlines a clear timeline for regulatory milestones from the submission of Substantial Equivalence (SE) applications to the establishment of the list of harmful and potentially harmful compounds in tobacco products. The agency has failed to meet many of these milestones and in turn granted extensions or enforcement discretion to the industry. The agency has also been delayed in the publication of rules and guidance documents related to regulatory submissions. These delays have presented a broad range of challenges to the industry and also to the organizations that service the industry including contract research laboratories (CROs). CROs are integral to the agency and the industry since they provide analytical testing services required for submission of product application. Since the passage of the TCA, CROs have prepared to meet the requirements of the FDA, often by trying to anticipate the requirements. Changes in FDA guidance has resulted in excess capacity or unneeded capabilities in the CROs. This presentation will focus on how CROs have responded to the challenges related to the changing regulatory landscape with a focus on the impact of changes and delays in tobacco product regulation guidance from the FDA.

Effect of the addition of SBA-15 in the pyrolysis of a reference tobacco under inert and oxidizing atmosphere

CALABUIG E.; MARCILLA A.; BELTRÁN M.I.
University of Alicante, San Vicente del Raspeig, Spain

The composition of evolved products from slow pyrolysis under inert and oxidizing atmospheres of a reference tobacco 3R4F from the Reference Cigarette Program of the College of Agriculture of the University of Kentucky and a mixture of the reference tobacco with a mesoporous catalyst SBA-15, with a fibre-like morphology, have been measured by using a multi-shot pyrolyzer, which was attached directly to a gas chromatography/mass spectrometry. The results obtained show that under inert atmosphere, the main decomposition rate occurs between 200-400 ºC with the generation of compounds characteristic of the decomposition of cellulose, hemicellulose, lignin and the generation of nicotine. The addition of SBA-15 to tobacco produces at a temperature below 300 ºC a general reduction of evolved gases, specifically decreasing in a greater proportion the formation of quinic acid, neophytadiene and nicotyrine. In contrast, at temperatures above 300 °C the opposite effect occurs and the generation of compound is greater, increasing the formation of acetaldehyde, chloromethane and aromatics. Under oxidizing atmosphere, at temperatures below 300 ºC, almost all compounds are reduced by adding SBA-15 to tobacco, such as glycerol, neophytadiene or nicotine. The combustion processes occur between 400-500 ºC with the generation of mainly CO2 and H2O and at temperatures above 300 ºC there is also a great reduction on the evolved gases like acid acetic, acetonitrile and aromatics.

Qualitative assessment of flavor ingredients in market electronic nicotine delivery systems

HUDSON A.G.(1); CULBERT C.G.(2)
(1) R.J. Reynolds Tobacco Company, Winston-Salem, NC, USA; (2) RAI Services Company, Winston-Salem, NC, USA

As the use of electronic nicotine delivery systems (ENDS) grows, it may be useful to understand the types of flavor compounds that are being added as ingredients to the ENDS currently available on the market. A qualitative analysis of flavor compounds was performed using gas chromatography mass/spectrometry (GC/MS) on the aerosols from 5 market ENDS products. Two cig-a-like, one tank, and two pod mod closed products were assessed with tobacco, menthol, and fruit flavored cartridges. The flavor compounds identified were sorted into 11 chemical categories. The qualitative results showed that the majority of flavor compounds found in the tobacco flavored cartridges were organic acids (pungent, sour notes), phenolic/pyrone compounds (smoky, sweet notes), and pyrazines/pyridines (nutty, earthy, cocoa notes). The menthol flavored cartridges primarily contained terpenoids (herbal, earthy notes) and menthol-related compounds. The most complex mixture of flavor compounds was found in the fruit flavored cartridges, with the majority being esters (aromatic, sweet notes), organic acids, and alcohols (sharp, grassy notes). Compounds identified in ENDS from all 11 chemical categories are also commonly present in traditional tobacco products. In general, these analyses revealed that similar types of flavor compounds are being used within the ENDS market.

Use of cilia beat frequency in an in vitro airway model (EpiAirway-FT) as a tool for assessing viability

WOODHAMS A.; KIDD D.; HOLLINGS M.
Covance, Harrogate, UK

Human respiratory tract cilia play an essential role in clearing mucus and debris from the airway (mucociliary clearance). This important defence mechanism process protects the lungs from aerosolised pathogens and pollutants and can be directly attributed to cilia beat frequency (CBF). The Sissons-Ammons Video Analysis (SAVA) system (Ammons Engineering, USA) is a combination of high-speed video, microscopy and image analysis which makes it possible to evaluate CBF in vitro. In a Proof-of-Concept study using EpiAirway™ tissues (Mattek Corporation, Ashland, USA) and a basolateral treatment regime, we optimised the CBF analysis parameters and recorded 2.6 second movies at 10x magnification. We then applied either chemicals with known ciliomotility effects (Adenosine triphosphate [ATP], Procaterol Hydrochloride [USAN], Hydrogen Peroxide [H2O2] or Benzalkonium Chloride [BC]) or adjusted the culture medium pH from 4 to 10 and analysed tissues after 30 minutes and 2 hours. Our results showed an untreated tissue average CBF of 17.62 Hz n=40. ATP (0.01mM) and USAN (0.01mM) increased CBF (as expected) to 19.68 Hz and 20.42 Hz respectively and BC (3mM) decreased CBF (as expected) to 2.5 Hz however H2O2 (0.1mM) caused an unexpected increase in CBF (19.55 Hz) even when tested at 10mM (19.37 Hz). From the pH testing, only pH4 adversely affected CBF yielding ciliostasis. To expand this work, EpiAirway™ AIR-200 tissues were exposed to 3R4F cigarettes (Health Canada Intense puffing regime) for 88 minutes using a Vitrocell® VC10® machine (VitroCell GmbH, Germany). CBF was slightly elevated at 2L/min compared to the air control and was undeterminable at higher concentrations. These data demonstrate CBF analysis could be useful for studying cytotoxic effects of chemicals, pH and aerosol exposure on airway tissue in vitro.

Toxicological assessment of e-liquid formulations using in vitro genotoxicity and cytotoxicity assays

COFFA B.G.; DOSHI U.; ZHANG Jingjie; McKINNEY W.J.; LEE K.M.
Altria Client Services LLC, Richmond, VA, USA

In the Electronic Nicotine Delivery Systems (ENDS) Premarket Tobacco Application (PMTA; 2016) Draft Guidance, the FDA recommends a full assessment of the toxicological profile associated with new tobacco products, using in vitro toxicology (e.g., genotoxicity and cytotoxicity) studies. As part of a toxicological hazard assessment, we tested flavor varieties of e-liquids used in MarkTen® e-vapor products (a total of 14 formulations) and two carrier formulations (propylene glycol, glycerin, with 0% or 5% nicotine) to a standard battery of in vitro cytotoxicity (Neutral Red Uptake [NRU]) and genotoxicity (Ames and micronucleus [MN]) assays according to OECD guidelines and using the maximum doses suggested for mixtures. The e-liquid formulations were characterized for key ingredients (propylene glycol, glycerin, and nicotine). All the formulations were non-cytotoxic per NRU assay (viability >80%). None of the e-liquids were mutagenic in Ames assay, however some reduction in background lawn was observed with carrier formulation at the high (5%) nicotine content. In the MN assay, 3/14 MarkTen® flavor formulations induced a weak but statistically significant increase in micronuclei formation, resulting in positive or equivocal findings according to OECD 487: All three flavor formulations were further evaluated in an in vivo combined genotoxicity (MN and Comet; OECD 474/489) assay and found to be negative for genotoxic endpoints. Therefore, consistent with International Conference on Harmonization (ICH) S2(R1) genotoxicity testing guideline (2012), the tested e-liquids were regarded as negative for genotoxicity under the conditions of the assays.

Evaluation of novel, oral tobacco-derived nicotine products for HPHCs

DANIELSON T.L.; McFARLANE C.B.; BROWN A.P.; JIN Xiaohong; WILKINSON C.T.; PITHAWALLA Y.B.; McKINNEY W.J. Jr.
Altria Client Services LLC, Richmond, VA, USA

VERVE® Discs and Chews are oral, non-dissolvable, tobacco-derived nicotine products. In May 2016, the U.S. Food and Drug Administration (FDA) issued a final rule to deem e-cigarettes, cigars and all other tobacco products, including VERVE® to be subject to the Federal Food, Drug, and Cosmetic Act (the FD&C Act), as amended by the Family Smoking Prevention and Tobacco Control Act (Tobacco Control Act). Manufacturers of regulated tobacco products are required to report to FDA quantities of Harmful and Potentially Harmful Constituents (HPHCs) by November 8, 2019. FDA has not issued specific guidance for reporting HPHCs for novel tobacco products, such as VERVE®, as they have for certain other regulated tobacco products. Absent specific guidance from FDA, we measured HPHCs in VERVE® according to the requirements for smokeless tobacco, recognizing that these products do not meet the statutory definition of a smokeless tobacco product. The objective of this work was to measure HPHCs in VERVE® Discs and Chews products and compare those results to other commercially available oral tobacco products and an oral nicotine replacement therapy (NRT) product. Results show the absence of detectable levels or significant reductions in HPHCs compared to other oral tobacco products and comparable HPHC results to the NRT.

Quantification of nicotine related impurities in novel, oral tobacco-derived nicotine products

DANIELSON T.L.; McFARLANE C.B.; NOE M.R.; PITHAWALLA Y.B.; McKINNEY W.J. Jr.
Altria Client Services LLC, Richmond, VA, USA

VERVE® Discs and Chews are oral, non-dissolvable, tobacco-derived nicotine products. In May 2016, the U.S. Food and Drug Administration (FDA) issued a final rule to deem e-cigarettes, cigars and all other tobacco products, such as VERVE®, to be subject to the Federal Food, Drug, and Cosmetic Act (the FD&C Act), as amended by the Family Smoking Prevention and Tobacco Control Act (Tobacco Control Act). As a result, these deemed tobacco products will ultimately require a market authorization from FDA through the premarket tobacco application (PMTA) pathway to either remain on the market or to enter interstate commerce. Section 910(b)(1)(B) of the FD&C Act requires "that your PMTA include a full statement of the properties of the new tobacco product. The 'full statement of the properties' of the new tobacco product should include a full narrative description of the tobacco product, including: Established shelf life of the product to include data establishing the stability of the product through the stated shelf life." The US and European Pharmacopeias recommend purity specifications for nicotine intended for pharmaceutical products; however, there are no official purity specifications recommended for the nicotine used in tobacco products. We developed a sensitive, selective, and robust analytical method (LC-MS) to quantify nicotine-related impurities listed in Pharmacopeia guidelines in VERVE® Discs and Chews and monitored these impurities over time. This selective and sensitive method provides data suitable for quantitative risk assessments and for stability studies.

Equivalence testing in demonstrating substantial equivalence for new tobacco products

FROST-PINEDA K.(1); CAMPBELL L.R.(1); POLSTER M.(2); CURTIN G.M.(1)
(1) RAI Services Company, Winston Salem, NC, USA; (2) Naxion, Philadelphia, PA, USA

Under Section 905j of the Tobacco Control Act, manufacturers must submit reports to the FDA to demonstrate that a new tobacco product is substantially equivalent (SE) to a "predicate product." Demonstrating equivalence with traditional statistical significance testing is, however, challenging for two reasons. First, the tests are specifically designed to demonstrate that two samples are different (i.e., to reject the null hypothesis that the samples are the same), and it is inappropriate to conclude equivalence when the null hypothesis cannot be rejected. Second, with very large samples, small differences can rise to statistical significance, but not actually be meaningful. As a result, there may be value in using equivalence testing, which examines whether or not the difference between means is smaller than a smallest effect size of interest by testing two null hypotheses (i.e., larger than the upper bound and smaller than the lower bound). If both hypotheses are rejected, the two means are essentially equivalent. Here we report on results of consumer testing in which likelihood of use, appeal, and risk perceptions were measured in a large sample (n=4,720) of current, former, and never tobacco users. The findings demonstrate that traditional statistical significance t-tests and equivalence tests generally yield similar results (i.e., significant differences are not equivalent), but that the two types of tests yield divergent results when: (a) sample sizes are very large and effect sizes are very small (i.e., significant differences that are not equivalent); or (b) sample sizes are small and effect sizes are large (i.e., differences are neither significant nor equivalent). Consideration is given to how to interpret discrepant findings in the context of SE product applications.

Method development for pH analysis of ends aerosol and e-liquid

HEFNER T.(1); ST. CHARLES F.K.(2); ANDERSON J.(3)
(1) R.J. Reynolds Tobacco Company, Winston Salem, NC, USA; (2) St.Charles Consultancy, Winston Salem, NC, USA; (3) Eurofins Lancaster Labs, Winston Salem, NC, USA

E-liquids are hygroscopic, non-aqueous liquids containing a small amount of water (~3-15%). When the pH of a non-aqueous liquid is measured using conventional methods and calibration buffers, the value can only be regarded as an "apparent pH". This is because the ionization constant of the acid or base, the dielectric constant of the medium, the liquid-junction potential, and the hydrogen-ion response of the glass electrode are all changed (paraphrased from USP 39 <791>). In addition, the hygroscopic nature of e-liquids results in uncontrolled variation in water content depending on the storage and measurement environment. This leads to an uncontrolled variation in the "apparent pH" value. Methodology was developed to measure the pH of e-liquids diluted with water to provide a more robust measurement which was compatible with conventional pH buffers and methods. To validate the dilution level, the pH of 25 commercial e-liquids was measured at dilution levels from 1:1 to 20:1 (w:w) water : e-liquid. The commercial samples covered a wide range of nicotine levels and glycerin to propylene glycol ratios. The most robust dilution level was found to be in the 7:1 to 9:1 dilution range. For aerosol measurement, an 8:1 dilution range was found to give more stable pH results. It was found that glass-fiber Cambridge filter pads (CFPs) used for aerosol collection could alter pH to higher values. Whatman QM-A quartz-fiber filter pads were found to have minimal influence on pH.

Genotoxicity evaluation of tobacco and nicotine delivery products: Part 1. Mouse lymphoma assay

LEVERETTE R.(1); BOMBICK B.(1); FOWLER K.(1); BREHENY D.(2); GACA M.(2); THORNE D.(2); CLEMENTS J.(3); LLOYD M.(3); McENANEY S.(3)
(1) RAI Services Company, Winston-Salem, NC, USA; (2) British American Tobacco (Investments) Ltd., Southampton, UK; (3) Covance Laboratories Ltd, Harrogate, UK

In vitro studies have been widely used to support the toxicological evaluation of chemicals and complex mixtures including cigarette smoke. In this study a variety of test matrices from different tobacco and nicotine delivery products was assessed against a Kentucky reference (3R4F) cigarette. The mouse lymphoma assay (MLA) is underpinned by OECD guideline 490 and ICH S2(R1) guidance and is a recognised in vitro genotoxicity test battery assay. The aim of this study was to assess the suitability of the MLA with a variety of tobacco and nicotine product test matrices. Testing was conducted in general accordance to OECD Guideline 490 and ICH S2 (R1) test guidance. The same samples were also assessed using the in vitro micronucleus assay; results are reported separately (Part 2). Total particulate matter (TPM) from a 3R4F cigarette was compared against pad-collected aerosol matter generated from a commercial electronic nicotine delivery system (ENDS), a commercial e-liquid, and TPM from a commercial tobacco-heating product (THP) using the in vitro MLA. Exposures were conducted for 3 hr ±S9 metabolic activation and 24hr -S9 conditions at concentrations up to 500 µg/mL. Under all treatment conditions, 3R4F produced a clear positive response with regard to induction of mutation. In contrast, no marked induction of mutation was observed for the e-liquid, ENDS and THP. Based on the results of this study, the mouse lymphoma assay can be used effectively for the assessment of these test matrices.

Genotoxicity evaluation of tobacco and nicotine delivery products: Part 2. In vitro micronucleus assay

LEVERETTE R.(1); BOMBICK B.(1); FOWLER K.(1); BREHENY D.(2); GACA M.(2); THORNE D.(2); CLEMENTS J.(3); McENANEY S.(3); WHITWELL J.(3)
(1) RAI Services Company, Winston-Salem, NC, USA; (2) British American Tobacco (Investments) Ltd., Southampton, UK; (3) Covance Laboratories Ltd, Harrogate, UK

In vitro studies have been widely used to support toxicological evaluations of chemicals and complex mixtures including cigarette smoke. In this study a variety of test matrices from tobacco and nicotine delivery products were assessed against a 3R4F reference cigarette using the in vitro micronucleus assay (IVMN). Assays were conducted in general accordance to OECD Guideline 487 and ICH S2 (R1) guidance. Samples were also assessed using the mouse lymphoma assay (Part 1). 3R4F total particulate matter (TPM) was first assessed with CHO, V79 (both rodent) and TK6 lymphoblastoid (human) cell lines with 3hr exposures ±S9 metabolic activation and extended -S9 treatments with/without a 1.5 - 2 cell cycle length recovery period at doses up to 500µg/mL. CHO, V79 and TK6 cells gave varied positive responses, with V79 being most responsive. The extended recovery treatment increased assay sensitivity for CHO and V79 cells; this was less clear in human TK6 cells. V79 cells were taken forward for further assessments. 3R4F TPM was compared against pad-collected aerosol matter generated from a commercial electronic nicotine delivery system (ENDS) , a commercial e-liquid, and TPM from a commercial tobacco-heating product (THP) using the same treatment schedules described above. Across all treatment regimens with V79 cells, clear negative responses were observed for the e-liquid, ENDS and THP samples, while 3R4F elicited a clear positive response. The most potent 3R4F responses were observed following extended treatment -S9 with recovery, suggesting this to be a more appropriate treatment schedule for the assessment of tobacco and nicotine product test matrices. Based on the results of this study the IVMN assay can be used effectively for the assessment of these test matrices.

Characterization of the Vitrocell® high throughput exposure modules using different tobacco product types

LEVERETTE R.(1); KEYSER B.(1); BOMBICK B.(1); HOLLINGS M.(2); SEYMOUR A.(2)
(1) RAI Services Company, Winston-Salem, NC, USA; (2) Covance Laboratories Ltd, Harrogate, UK

The development of whole smoke/aerosol exposure systems provides a means to conduct in vitro assessment of freshly generated whole smoke and aerosol from combustible and tobacco heating products (THP) as well as electronic nicotine delivery systems (ENDS). One challenge with such systems is ensuring sufficient throughput for in vitro toxicological studies in a timely manner. Vitrocell® has developed a high throughput whole smoke/aerosol exposure module designed to deliver concurrently up to seven different doses of smoke/aerosol and a clean air control to 48 wells of bacterial or mammalian cell cultures (six wells per dose). Characterization of this system was conducted with a series of experiments designed to assess smoke/aerosol delivery and biological responses from a Kentucky Reference 3R4F combustible cigarette or a commercially available THP. Dilution airflows consisting of 0.5 – 10 L/min for 3R4F and 0 (undiluted) - 4 L/min for the THP were evaluated. Smoke/aerosol deposition was quantified using fluorescence measurements (Ex 355/Em 485) of captured particulate matter and chemical analysis (e.g., glycerol, nicotine) of either DMSO (3R4F) or PBS (THP) traps within the module. Further characterization of the high throughput module was performed with the Neutral Red Uptake (NRU) assay to determine the cytotoxic response to 3R4F whole smoke. Current results demonstrate a dose-dependent deposition of smoke/aerosol constituents and a characteristic dose-dependent decrease in cell viability as indicated by the NRU assay.

Development of a 3D human oral tissue model for oral permeation of smokeless moist snuff

KEYSER B.(1); ROWE J.(2); WEIDMAN R.(2); FIELDS W.(1)
(1) RAI Services Company, Winston-Salem, NC, USA; (2) R.J. Reynolds Tobacco Company, Winston-Salem, NC, USA

Tobacco product use is a risk factor in the development of oral cancer, although epidemiology studies show this risk is far less with smokeless tobacco product use than cigarette smoking. While smokeless tobacco contains harmful and potentially harmful constituents (HPHCs), the oral permeation of HPHCs is not completely understood. The goal of this study was to determine the suitability of a 3D human buccal oral tissue model, EpiOral™, to study this process by using an extract of CORESTA moist snuff smokeless tobacco reference (CRP2). CRP2 extract was prepared in complete artificial saliva (CAS) with a ratio of 300 mg of CRP2 to 1 mL of CAS, sterile filtered and stored at -80°C until time of exposure with the stock extract or a 150 mg/mL dilution. CRP2 extract (150 or 300 mg/mL), CAS alone (vehicle control) or caffeine/CAS solution (0.15%; positive control) were applied to the apical side of EpiOral™ tissues and permeation was measured for 3 hours with fresh receiver solution (basolateral compartment) replaced every 30 minutes. After the last time point, EpiOral™ tissues were rinsed and cytotoxicity (via 3-(4,5-d-2-yl)-2,5-diphenyltetrazolium bromide; MTT) was measured. Subsequently, tobacco-specific nitrosamines (TSNAs) and caffeine concentrations were measured in each receiver solution by liquid chromatography (LC). Application of CRP2 extract (150 or 300 mg/mL), CAS alone or caffeine/CAS solution (0.15%) to the EpiOral™ tissues did not produce cytotoxicity as measured via MTT. Permeation over time (i.e., flux) was observed to achieve a linear steady state for each analyte measured over the 3 hours. Collectively, these data suggest that the EpiOral™ model could increase the understanding of the permeation of different tobacco constituents from moist snuff smokeless tobacco into oral tissue.

Development of a small airway lung model using microfluidic lung-on-a-chip device

MAKENA P.(1); PRASAD G.L.(1); GORDON D.(2); HAITHCOCK D.(2); PRABHAKARPANDIAN B.(2); PANT K.(2)
(1) RAI Services Company, Winston-Salem, NC, USA; (2) SynVivo Inc., Huntsville, AL, USA

Current in vitro models of small airway lung, which have primarily been developed on a transwell membrane with epithelial cells cultured at air liquid interface (ALI), lack certain elements of the in vivo physiology such as fluidic shear and endothelial cell component. We have developed a small airway lung-on-a-chip model that mimics the small airway architecture and physiology. The device is comprised of a microfluidic chip, commercially available human bronchial tissue sample-derived epithelial cells and maintains ALI across the epithelial cells. Additionally, differentiated epithelium is surrounded by human primary micro vascular endothelial cells. During method development phase, several culture conditions including timing of ALI growth of epithelial cells with endothelial cultures, co-culture media, and cell seeding density were optimized. Cells were differentiated into small airway architecture for 14 days and were monitored up to 4 weeks. The epithelial phenotypic assessments show presence of mucus production, goblet cells, ciliated cells, and formation of tight junctions. Several assays were developed to evaluate cytotoxic endpoints including cell viability, oxidative stress (ROS), cell death (TUNEL), and biochemical measures such as ICAM-1, epithelial-endothelial barrier integrity and permeability. Our results demonstrate a significant 1) decrease in cell viability when exposed to ethanol, 2) increase in ROS, ICAM expression, and epithelial permeability observed on treatment with TNF-α/IFN-γ, and 3) increase in apoptosis when treated with DNase I. This small airway lung model will provide useful information in understanding the interactions of toxicants and biomolecules and offers a physiologically relevant in vitro model to assess local lung effects caused by various toxicants including tobacco product emissions.

Microfluidic small airway lung model for toxicity screening of tobacco products

MAKENA P.(1); PRASAD G.L.(1); GORDON D.(2); HAITHCOCK D.(2); PRABHAKARPANDIAN B.(2); PANT K.(2)
(1) RAI Services Company, Winston-Salem, NC, USA; (2) SynVivo Inc., Huntsville, AL, USA

Physiologically relevant in vitro lung models are valuable tools to accurately assess the cytotoxicity of inhaled toxicants. Epithelial cells cultured in transwells do not accurately represent the cellular architecture, physiology, and fluidic shear present in vivo. Recent advances in microfluidics have led to the development of bioengineered, three-dimensional human small airway models composed of human bronchial epithelial (NHBE) cells surrounded by human lung microvascular endothelial (HMVE) cells. In this study, we evaluated the potential of a human small airway lung-on-a-chip model to assess cytotoxicity and functional endpoints following exposure to cigarette total particulate matter (TPM). The microfluidic device is comprised of a central air channel and lateral vascular channels separated by porous architecture. Using an optimized co-culture protocol, NHBE and HMVE cells were co-cultured using a combination of air and fluidic pumps. The cells were exposed apically to different concentrations of TPM for 4 hours and 1) cell viability, 2) oxidative stress, 3) cell death, and 4) epithelial permeability were assessed after 20 hours post exposure. TPM induced a dose-dependent decrease in cell viability and increased oxidative stress, cell death, and epithelial permeability. These results 1) show the potential of human small airway in vitro lung model to mimic physiologically relevant conditions and 2) lay a foundation to study the effects of different tobacco products. This model enables our understanding of key molecular events that lead to early biochemical, toxicological, and/or physiological perturbations and will be useful in identifying relevant biomarkers of effect.

Mechanisms for WSCM induced cytotoxicity to human aortic endothelial cells

MAKWANA O.(1); FLOCKTON H.(2); SMITH G.(2); WATTERS G.(2); NISAR R.(2); FIELDS W.(1); BOMBICK B.(1)
(1) RAI Services Company, Winston-Salem, NC, USA; (2) Covance Laboratories Ltd, Harrogate, UK

Chronic exposure to cigarette smoke can lead to endothelial dysfunction and potentially endothelial cell death, which is often associated with atherosclerosis. Here, we used Human Aortic Endothelial Cells (HAECs) from four healthy donors to determine smoke-induced cytotoxicity. HAECs were exposed to whole smoke conditioned media (WSCM) generated from 3R4F reference cigarettes over a range of 0-8000 ng/mL nicotine equivalence (n.e.) concentrations. Cytotoxicity was evaluated 24 hours post-exposure via neutral red uptake (NRU) and/or reduction of 3-(4,5-dimethyl-2-thiazolyl)-2,5-diphenyl-2H-tetrazolium bromide (MTT) to formazan for each exposure concentration and compared to vehicle control. The IC50 was then calculated from the concentration-response. Similar IC50 values were observed in both the NRU and MTT assays for the four donors across the dose range of 0-8000 ng/mL n.e. of WSCM. To examine the mechanism responsible for WSCM-induced cytotoxicity in HAECs at varying n.e. concentrations, necrosis (propidium iodide: PI) and apoptosis (Annexin V) markers were assessed via flow cytometry. Annexin V-positive cell populations increased in a dose-dependent manner while increases in PI-positive cell populations only occurred at the highest doses of WSCM (5000-8000 ng/mL n.e.). To further investigate the induction of apoptotic cells after WSCM exposure, we conducted Western blotting experimentation using caspase-3 cleavage as a marker for apoptosis. Western blots confirmed that apoptosis occurs at the higher concentrations of WSCM which coincide with reduced HAEC survival. We conclude that WSCM produces comparable cytotoxicity in vitro when applied to HAECs from four donors as assessed using both the NRU and MTT assays. Additionally, the mechanism of toxicity appears to be dose-dependent and conserved among the four donors as assessed by distinct measures for apoptosis.

Non-targeted analysis for differential screening of tobacco samples

MILLER J.H. IV; PITHAWALLA Y.B.; DANIELSON T.L.; XU Dongmei
Altria Client Services LLC, Richmond, VA, USA

We have developed a non-targeted liquid chromatography mass spectrometry method for the analysis of tobacco samples which is able to differentiate between types of tobacco and tobacco leaf blends. Non-targeted analysis of samples has broad applications across the food industry and in the evaluation of natural products. For example, this approach has been used by the food industry to evaluate purity and quality of fruit juices, olive oil and tea. Our lab explored applying a non-targeted analysis using LC/MS-TOF to differentiate between types of tobacco and tobacco leaf blends. This work can provide valuable information to improve our understanding of differences in types of tobacco, curing processes, tobacco origin and the impact of leaf blending. In addition, it may help us identify compounds that contribute to a tobacco's key characteristics. We used a Waters Acquity UPLC® with Waters Premier Quadrupole Time of Flight (QTOF) mass spectrometer with positive electrospray ionization in full scan mode. Chromatographic separation was conducted using a reverse phase gradient with a BEH C18, 2.1 x 100mm 1.7 µm column from 5% Acetonitrile/95% Ammonium Formate to 95%Acetonitrile/5% Ammonium Formate over 16 minutes. The Waters MarkerLynx XS software was able to find hundreds of marker compounds in a variety of tobacco samples. Additional data processing in the software provided principal component analysis for the markers, and grouped the tobacco samples in scores plots. Loadings plots provided insights as to which markers are responsible for the unique groupings. MarkerLynx XS software allowed for differentiation of tobacco types and leaf blends. Future work to identify key markers compounds using accurate mass analysis could provide insights into the differences in chemical composition of these tobacco samples.

Thermal profiling methods for the evaluation of electronic nicotine delivery systems (ENDS)

ROGERS R.(1); SZABO D.T.(1); STRASSER J.(1); REYNOLDS C.(2)
(1) RAI Services Company, Winston-Salem, NC, USA; (2) R.J. Reynolds Tobacco Company, Winston-Salem, NC, USA

Electronic nicotine delivery systems (ENDS) contain tobacco-derived nicotine formulated in a liquid ("e-liquid") typically consisting of glycerin, propylene glycol, water and flavoring ingredients. The e-liquid is aerosolized via a heating element in the cartridge, containing the e-liquid, powered by a battery in the power unit. In the 2016 US Food and Drug Administration (FDA) draft guidance "Premarket Tobacco Product Applications for Electronic Nicotine Delivery Systems", the FDA recommends a description of the heating temperature of an ENDS in order to characterize the aerosolizing apparatus. Thermal profiling studies measure the exposed surface temperatures of various cartridge component locations during aerosol generation. Thermocouples and thermal imagers, being used simultaneously, allow for temperature measurements of the heating element and other structural materials within the cartridge. A methodology was developed to measure ENDS temperature profiles during each puff of the product using thermocouples attached to the cartridge materials and an infrared camera to measure temperatures of the heating element. As different ENDS contain different designs and heating element materials and locations, this method contains flexibility to adequately measure temperatures across a range of ENDS. This thermal profiling methodology allows for temperature measurements of the aerosolizing apparatus during use, and provides heating temperature information in order to characterize the product, as recommended per FDA's draft guidance.

A novel approach to the assessment of electronic cigarette topography characteristics

SMITH J.S.(1); ALDERMAN S.(2)
(1) RAI Services Company, Winston-Salem, NC, USA; (2) RAI Innovations Company, Winston-Salem, NC, USA

The use of electronic nicotine delivery systems (ENDS) has increased dramatically over the last decade. However, during this same period of time, there has been very little empirical evidence to describe how subjects use these new products. One reason for this scarcity of data is that techniques and approaches used to collect topography (puff) characteristics have yet to catch up with the changes in format and function of new ENDS products. We have developed a new approach that will assess ENDS topography use characteristics in both controlled (in-clinic) and naturalistic (ambulatory) environments. The Product Use and Behavior (PUB) instrument is positioned between the power unit and cartridge of an ENDS product and acts as a sentinel, recording the periods of product activation over a continuous period of use without impacting how the subject interacts with the ENDS product. This new approach provides data on puff duration, inter-puff interval, total number of puffs, battery characteristics (voltage/current), as well as how these characteristics vary across use sessions in the form of cumulative use patterns over time. Additionally, data captured by the PUB instrument can also be used to estimate mouth level exposure (eMLE) per puff and over product use periods. The PUB instrument produces data that will provide insights into how subjects interact and use ENDS products.

Low cost electrochemical sensor as a means for determining dry wick conditions in tank ends devices

TINDALL I.; MASON T.
Cerulean, Milton Keynes, UK

Previous experiments using FTIR analysis of the aerosol produced by e-cigarettes in "dry wicking" conditions has shown trace levels of CO accompany the production of carbonyls such as Formaldehyde. This may be the result of wick scorching or the catalytic decomposition of the residual e-liquid. The study was designed to explore the conditions under which CO is generated from ENDs devices and the suitability of low cost CO detectors for this purpose. A vaping system was equipped with a 0-1000ppm electrochemical sensor in a simple exposure chamber fitted with a 22 micron PTFE hydrophobic barrier. Data was logged at 1 second intervals. 13 different ENDs devices were tested to exhaustion or with limited e-juice. Cigalike systems produced no CO before reaching a "non-vaping" condition. The exception was one of the menthol products that showed some CO although this is almost certainly a cross interference for the sensor caused by the menthol. Significant CO spiking (350ppm ->1000ppm) was observed for exhausted tank systems although one system fitted with a glass fibre wick did not show spiking even when fully dry. Heat not burn systems (HNB) consistently delivered approximately 100 – 200 ppm CO per puff (depending on system) as measured by the system deployed. The simple electrochemical sensing device is sufficiently sensitive to detect changes in Co yield indicative of vaping through "dry wicking". The most suitable location for the sensor was found to be in the exhaust line of the puff engine. The wick and coil arrangement of tank systems appears significant in the production of CO at end of life, and CO can be used as a dry wicking indicator for specific products.

Application and use of e-cigarettes HPHC methods for the analysis of heated tobacco products

GILLMAN I.G.; HUMPHRIES K.; MARTIN A.; LASSITER S.; HILLDRUP J.
Enthalpy Analytical LLC, Durham, NC, USA

The interest in e-cigarette research has resulted in the development of methods for quantitative analysis of chemical constituents present in their aerosols. The emissions of e-cigarettes contain not only nicotine, propylene glycol, glycerin, water and flavours, but also nicotine related impurities and thermal degradation products. We have developed and validated analytical methods for the analysis of e-cigarette aerosols including the compounds listed in the US FDA Draft PMTA guidance document. These include primary constitutes, aldehydes, VOCs, organic acids, metals, aromatic amines, nicotine related impurities and benzo[a]pyrene. An emerging tobacco product; Heated Tobacco Products (HTPs) produce an aerosol by heating of tobacco. Since tobacco is not combusted in these products, the majority of the resulting aerosol is comprised of water, humectants and volatile compounds. Given the similarities between e-cigarettes and HTPs, we hypothesized that e-cigarettes methods would also be suitable for analysis of HTPs emissions. In this study, we have applied our e-cigarette methods to the analysis of the iQOS HEETS HTPs. The aerosols were collected under intense smoking condition, using a device holder designed for e-cigarettes, and analyzed for HPHCs. Additionally, the aerosols were analyzed for water using both the ISO method for cigarettes and an in-house developed method. We found that methods developed and validated for e-cigarettes could be successfully used for the analysis of HTPs. With the exception of nicotine and water, the emissions from HTPs were found to be more similar in emissions to e-cigarettes aerosol than cigarettes smoke. We found that e-cigarette methods are well suited for the analysis of HTPs due to the lower calibration ranges and selected compound list as compared to methods developed for conventional cigarettes. Data will be reported for all the FDA PMTA compounds in the HTPs, including comparisons of expected ranges in both e-cigarettes and cigarette smoke.

Benchmark dose modeling of genomic data to assess the functional effects of cigarette smoke exposure on peripheral blood mononuclear cells

LIU Gang; TRAN Q.T.; PRASAD G.L.
RAI Services Company, Winston Salem, NC, USA

Benchmark dose (BMD) modeling of toxicogenomics data has gained interest from regulatory agencies for chemical risk and toxicity assessment. To date, limited studies have used this tool for assessing toxicological effects associated with the use of tobacco products. In this study, we applied an integrative approach combining BMD modeling, gene ontology (GO) classification, and gene set enrichment analysis to assess dose-dependent microarray data for quantitative assessment of the biological effects of cigarette smoke exposure. Gene expression changes in peripheral blood mononuclear cells upon exposure to three nicotine equivalent doses of whole smoke-conditioned medium were used as a case study. Our approach consists of three steps. First, "BMDExpress" software was used to model gene expression data, and the BMDs of every dose-responsive genes and their associated GO categories were estimated. Second, dose-sensitive biological pathways were identified based on their BMD values and the percentage of differentially expressed genes in the pathway. Third, dose-response gene set enrichment analyses were performed to identify the key pathways that are significantly enriched with genes showing significant monotonic dose response. Our analysis identified 64 dose-sensitive pathways, which are mainly associated with three key cellular processes including inflammatory response, cellular response to stress, and Toll-like receptor signaling. For example, a biological pathway named "response to nitrosative stress" was observed to be dose-dependent and its BMD was estimated. In addition, our enrichment analysis identified seven significantly enriched pathways such as amino acid transport across the plasma membrane and small cell lung cancer pathway. Together, we present a case study using BMD modeling to identify dose-responsive genes and pathways associated with smoke exposure, and calculate the reference doses at which cellular processes are altered.

Effects of smoking on seminal cytokine network

OMU A.E.; OMU F.E.
University of Kuwait, Safat, Kuwait

Introduction: The seminal cytokine network is involved in priming female reproductive tractreception for embryo implantation and may have a role in infertility and can be affected by smoking tobacco. Objective; To evaluate the association between smoking tobacco and sperm parameters and the effect on seminal cytokine network. Subjects and method; Eighty men who presented with infertility were evaluated. Clinical evaluation including smoking habits were noted and semen analysis was carried out according to WHO guidelines. A panel of cytokines TNF-α, IL-12, IL-6,IL-13, IL10 and IL-4 were estimated in smokers and non-smokers using Enzyme-linked immune sorbent assay (ELIZA). After a program of smoking cessation, the seminal cytokines were repeated in those men with significant cessation of smoking through low carbon monoxide level in their breath using edfont smokerlyzer. Result; Of the 80 men, 29 (36%) were addictive smokers and 51 (64% were non-smokers. The T helper proinflammatory cytokines expression were significantly higher in smokers IL-6; 36 ± 8 versus 26 ±4, p<0.05, IL-12, 44±6 versus 27±6, p <0.01, TNF α=12 ±3 versus 7 ± 1.3, p <0.05. On the other hand, T helper 2 expression was higher in non-smokers than smokers. IL-13; 12±3 versus 6±2, p<0.01, IL-10: 9±4 versus 4±1.2, p<0.01, IL-4; 6± 1.2, versus 3±1.2, p<0.05. Semen analysis revealed Normozoospermia, Oligozoospermia, Asthenoospermia and Leucocytospermia, had higher experssion of proinflammatory cytokines IL-12, IL-6 and TNF-α. Conclusion; Smoking is associated with proinflammatory cytokine (T helper 1) expression. This may explain the detrimental effect of smoking on sperm parameters and function.

Analysis of trimethylamine in mainstream cigarette smoke

ZHU J.; BROOKS C.; BOYD N.; TAYYARAH R.
ITG Brands LLC, Greensboro NC, USA

Trimethylamine (TMA) is a small volatile amine present in cigarette smoke at low levels as a known combustion by-product. It is hydrophilic in nature and exists in pure form as gas at room temperature. Because of these characteristics, its relatively low concentrations in cigarette smoke, and its low molecular weight, this constituent presents a challenge for quantitation for cigarette smoke characterization. Methodology in the literature for trimethylamine in cigarette smoke as a matrix is limited. In order to develop a robust quantitative method, several platforms and extraction techniques were explored. We would like to present our novel approaches using different platforms such as GC-NPD, GC-MS, GC-MS/MS to achieve targets for sensitivity and selectivity, and to discuss limitations of the platforms and how their results confirm each other.

Analysis of furfural in mainstream cigarette smoke by GC-MS

ZHU J.; BROOKS C.; ZICH D.; BOYD N.
ITG Brands LLC, Greensboro, NC, USA

Furfural is a natural aroma component in tobacco and is also present in cigarette smoke at low level. Methods exist in the literature for tobacco and for e-cigarette aerosol. On the other hand, published literature is limited for cigarette smoke analysis. Analysis of this constituent may be of interest for tobacco-related product characterization. Herein we wish to present the GC-MS method we have developed to quantitate furfural in cigarette smoke. An existing in-house smoking procedure using impingers for volatile organics was adapted with furfural-d4 added as an internal standard. The calibration range covers furfural levels in typical cigarettes including 3R4F and 1R6F cigarettes and lower tar cigarettes. Furfural was determined at approximately 1.4 µg/cig and 12 µg/cig for 3R4F smoked under ISO and HCI, respectively. Good recoveries were obtained on 3R4F cigarettes.

Comparisons of dependence on cigarettes and e-cigarettes: data from the path study

KIM M.M.(1); SHIFFMAN S.(2); SEMBOWER M.(2)
(1) RAI Services Company, Winston-Salem, NC, USA; (2) PinneyAssociates, Inc., Pittsburgh, PA, USA

An Item-Response-Theory-based dependence scale, developed on PATH data, assesses e-cigarette dependence, and for the first time allows for direct comparison of dependence across products. Data were from adult established users of e-cigarettes (ECIG) and/or cigarettes (SMOK; and no other tobacco) in PATH (Wave 1), a representative survey of the US. Dependence on SMOK and ECIG was contrasted across groups defined by history of use of each product (current users or quit in past year, N=7,625). Among established users of both SMOK and ECIG, dependence was much higher on SMOK; this was true both among current (SMOK: mean=3.11 [SE=0.05] vs. ECIG =1.61 [0.04], $p<0.0001$) and former users (SMOK =1.59 [0.12] vs. ECIG =1.11 [0.04], $p=0.0001$). Current smokers who reported currently using ECIG were more SMOK-dependent than never-users of ECIG (ECIG =3.07 [0.03] vs. never ECIG =2.69 [0.02], $p<0.0001$). Among current ECIG users, those who had quit smoking in the past year reported greater ECIG dependence than those who were still smoking, but their ECIG dependence was significantly lower than the SMOK dependence of current smokers, whether they currently used ECIG, had quit ECIG, or never used ECIG. In direct within-person comparisons between cigarettes and e-cigarettes, e-cigarettes were associated with less dependence. Greater cigarette dependence was seen in smokers who used e-cigarettes than in those who did not; more dependent smokers may gravitate to and stay with e-cigarettes. The highest level of e-cigarette dependence was seen in those who had quit smoking in the past year; heavier e-cigarette use is associated with smoking cessation. These people may have transferred their nicotine dependence from cigarettes to e-cigarettes, as hoped for by harm-reduction advocates.

Use patterns and tobacco use histories among users of diverse vapor products: cigalikes, tanks, and other vapor products

KIM M.M.(1); SHIFFMAN S.(2); SEMBOWER M.(2)
(1) RAI Services Company, Winston-Salem, NC, USA; (2) PinneyAssociates, Inc., Pittsburgh, PA, USA

There are several types of vapor products, but little research has characterized differences in their use or users. We examined the differences in use and characteristics of users of cigalikes, tanks, and other, 'variant' vapor products in a national online survey of US adults. A total of 11,601 past-30-day users of vapor products were identified in 2015-16 in the National Tobacco Behavior Monitor. Demographics, use patterns, and tobacco-use profiles were compared between exclusive and multiple device users. Exclusive use of cigalikes was most common (45%); exclusive tank use (23%) and concurrent tank and cigalike use (17%) were also common, and exclusive use of variant products least common (5%). User profiles differed significantly by product type. Cigalike and multiple device users tended to be male (60%, 62%, respectively). Users of variants tended to be Black (15%) or Hispanic (25%), whereas users of cigalikes (63%), tanks (70%), or both (61%) tended to be White. Tank users had the highest rate of daily use (36%), and also were the most likely to be previously-established smokers who had quit (27% of tank users), especially if they used tanks daily (55% were former smokers). Users of multiple vapor products were least likely to be former smokers (10%) unlike users of variant vapor products who were most likely to be infrequent users (42% using less than 5 days/month), and least likely to have a history of established tobacco use (78%). It is important to understand these differences in product selection; research should avoid combining various vapor products or assuming their users or patterns of use are homogeneous. Longitudinal data may further inform whether these patterns reflect migration between vapor products.

Determination of indogenous and leachable metals by ICPMS

STOGNER D.C.; STEVENSON B.; GRAY J.; BENNER D.N.; ADAMS C.; CHOUCHANE S.
Eurofins Lancaster Laboratories PSS, Winston-Salem, NC, USA

There is a need for measurement of low levels of both indogenous and leachable elements when evaluating electronic cigarette matrices. A method was developed and validated to accurately quantitate these analytes. The electronic cigarettes are smoked on a Cerulan 450e Linear Smoking machine and the aerosol is trapped in impingers containing nitric acid/purified water. The samples are analyzed on a Perkin Elmer 300 S ICPMS using normal mode, reaction mode and collision mode to ensure any interferences are removed. The method validation results will be presented.

Water analysis by volumetric Karl Fischer titration using direct heating and transfer to the Karl Fischer titrator

STOGNER D.C.; BENNER D.N.; STEVENSON B.; ADAMS C.; CHOUCHANE S.
Eurofins Lancaster Laboratories PSS, Winston-Salem, NC, USA

Tobacco and tobacco related products typically have been analyzed for water content by extracting into anhydrous methanol and then manually injecting into the Karl Fischer titrator. An alternate method has been developed whereby a sample is weighed into a sealed vial, and an automated oven is utilized to heat the tobacco and tobacco related products and drive the water to the Karl Fischer titrator using dry nitrogen. This method eliminates many possible interferences and is also applicable to electronic cigarette liquids. The method validation results will be reported.

Influence of the time of synthesis of SBA-15 on its effect as a catalyst reducing toxicant concentration in tobacco smoke

JUÁREZ-SERRANO N.; MARCILLA A.; BELTRÁN M.I.
University of Alicante, San Vicente del Raspeig, Spain

In recent years, SBA-15 has been suggested as an additive for reduction of toxic compounds in tobacco smoke. The structure of this type of materials is strongly dependent on the synthesis conditions. The objective of the present work is to analyze the influence of the time of the hydrothermal treatment on the structure of the obtained SBA-15 and its effect in reducing toxicants of tobacco smoke. SBA-15 were prepared according a well-known procedure and varying the time of synthesis. The materials were characterized by nitrogen adsorption isotherms, scanning electron micrographs (SEM), X-ray powder diffraction (XRD) and apparent density. Thermogravimetry was run on samples before and after hydrothermal treatment. 3R4F tobacco and catalyst mixtures were smoked under ISO 3308:2000 regime. The composition of the total particular matter (TPM) and the gas fraction smoke was quantified by GC-MS and GC-FID, respectively. Increasing the time of the hydrothermal treatment leads to a decrease of the apparent density, and improvement of the adsorption capacity of the materials obtained, as well as marked reductions in the concentration of different toxicants and nicotine in the smoke generated when mixing the catalysts with the 3R4F reference tobacco under ISO conditions. The effect is saturating and times of the hydrothermal treatment longer than 15 h lead to marginal improvements.

Determination of select leachable compounds in electronic cigarette aerosol and liquid by LCMSMS

RASHMAWI K.J.; SPANN N.; PORTER A.; ADAMS C.; CHOUCHANE S.
Eurofins Lancaster Laboratories PSS, Winston-Salem, NC, USA

Polymeric materials used in the manufacturing of electronic cigarette cartridges may contain compounds that have the potential to leach into the liquid or the aerosol of electronic cigarettes. A method was developed and validated for the determination of select leachable compounds such as bisphenol A, erucamide, Span 60, Irganox 1076, oxidized Irgafos 168, PET cyclic trimer in electronic cigarette liquid and aerosol using solvent extraction and LCMSMS. Samples are extracted with isopropyl alcohol and quantified based on the individual compound MRM transitions. The method validation results will be reported.

Trends in US smokers' perceptions of the relative risks of non-combustible tobacco products versus cigarettes

CURTIN G.M.(1); SHIFFMAN S.(2); ROHAY J.(2); PYPE S.(2)
(1) RAI Services Company, Winston-Salem, NC, USA; (2) PinneyAssociates, Inc., Pittsburgh, PA, USA

Smokers can benefit from switching to less harmful tobacco products, which requires knowledge of relative product risk. However, publicly shared information often emphasizes absolute rather than relative risks. To understand smokers' perceptions of product risk, we assessed perceptions of various non-combustible products, relative to cigarette smoking, among 31,269 US adult smokers in a series of cross-sectional surveys (2009-2017). Smokers aware of a product rated that product's risk relative to cigarettes (7-point scale, from "a lot more risky" to "a lot less risky"). Products included snus, other smokeless tobacco, tobacco-heating products, and vapor products (cigalikes, tanks, other vapor products). Logistic regression was used to analyze linear trends among the proportion of smokers who considered each product to be at least as risky as smoking. Analyses also compared risk perceptions of different vapor products. Generally, smokers' perceptions that non-combustible products are at least as risky as cigarettes increased significantly over time. This was true for tobacco-heating products (61% to 66%), cigalikes (39% to 44%), and tanks (43% to 48%); there was no change for other vapor products (45% to 47%) or snus (73% to 73%). Perceived relative risk of other smokeless tobacco declined significantly over time (83% to 80%). Among vapor products, tanks were the most likely and cigalikes the least likely to be rated at least as risky as cigarettes. Majorities, or near majorities, of smokers perceived each of the non-combustible products to be at least as risky as cigarettes. Except for smokeless tobacco, which large majorities deemed at least as risky as smoking, this misperception increased over time. Smokers' misunderstanding of the relative risks of non-combustible tobacco products likely impede harm reduction.

Multi-dimensional tipping point analyses: assessing simultaneous shifts in tobacco use patterns from a higher to a lower risk product

CURTIN G.M.(1); BACHAND A.(2); SULSKY S.(2)
(1) RAI Services Company, Winston-Salem, NC, USA; (2) Ramboll US Corporation, Amherst, MA, USA

The Dynamic Population Modeler (DPM(+1)) employs a birth cohort framework to estimate effects on population mortality if tobacco use patterns shift from a higher- to lower-risk product. DPM(+1) allows for evaluation of changes in use patterns within the context of 'tipping point' analyses, which estimate the magnitude of a beneficial use pattern needed to offset the population effect of harmful use patterns. In a birth cohort followed from age 13 to age 72, in 5-year intervals, we specified transition probabilities for a counterfactual scenario whereby 3% of individuals who would have never used tobacco instead initiate modified-risk tobacco product (MRTP) use (up to age 27; additional initiation), and 25% of those MRTP initiators transition to cigarette use (next age interval; gateway effect). Assuming a 92% reduction in all-cause mortality risk for the MRTP (relative to cigarettes), population harm was offset if at least 1.6% of smokers who would have continued to smoke instead switched to MRTP use (each age interval, for ages 18+; switching). In addition, ranges of probabilities for up to three transitions can be assessed simultaneously. Assuming 1-10%, 0-50% and 0-10% for additional initiation, gateway effect and switching, respectively, 2% switching offset the population harm resulting from, for example, 4% additional initiation with 20% gateway effect, 5% additional initiation with 10% gateway effect, and 6% additional initiation with 4% gateway effect. Tipping point analyses allow regulators to assess the magnitude of simultaneous changes in use patterns likely to result in an overall population benefit or harm. Such analyses may reduce the immediate need for empirical projections of beneficial and/or harmful changes in use patterns during regulatory decision making.

Comparing direct and indirect assessments of perceptions of reduced risk for a modified-risk tobacco product

SHIFFMAN S.(1); CURTIN G.M.(2)
(1) PinneyAssociates, Pittsburgh, PA, USA; (2) RAI Services Company, Winston-Salem, NC, USA

Manufacturers seeking to make claims about modified-risk tobacco products (MRTPs) must assess consumers' perceptions of absolute and relative risks for that product. We cross-validate data from two methods: (a) indirect assessments based on the difference between absolute risk ratings for the MRTP and cigarettes; and, (b) direct assessments comparing the MRTP to cigarettes. We also examine indirect assessments of consumers who, on direct assessment, indicate they don't know the risk of the MRTP relative to cigarettes. US adult current, former and never tobacco users (n=18,234) recruited from online research panels were shown advertisements stating that smokers who switch completely to Camel Snus (CS) could reduce their risk of lung cancer and respiratory disease. Respondents rated the absolute risks of CS and cigarettes, separately, on a 1-7 scale, and directly compared the risks of CS versus cigarettes for each disease. For lung cancer, the mean difference in absolute ratings for CS and cigarettes was -0.4±0.02 for those who indicated no difference in risk, based on direct assessment; -2.1±0.02 for those who indicated a lower risk for CS; -3.4±0.04 for those who indicated CS presented no risk at all; and, -1.0±0.04 for those who responded "don't know." Results were similar for respiratory disease. The tight correlation between direct and indirect modes of assessing perceived risk indicates reliability and validity for both methods. That respondents who indicated "don't know" in the direct assessment provided absolute ratings intermediate between 'same as smoking' and 'less risk' indicates their perceptions are very conservative; they do not believe CS is completely safe. The findings confirm the validity of these approaches for assessing perceived risks of MRTPs.

Consumers' perceptions of disease-specific modified-risk claims are best evaluated in a disease-specific manner

SHIFFMAN S.(1); CURTIN G.M.(2)
(1) PinneyAssociates, Pittsburgh, PA, USA; (2) RAI Services Company, Winston-Salem, NC, USA

Manufacturers seeking to make claims about modified-risk tobacco products (MRTPs) must assess consumers' perceptions of risk reduction for that product. Claims made for the MRTP and assessment of perceived risk reduction can be global (e.g., "reduced harm" and assessment of a broad spectrum of risks) or disease-specific (e.g., "reduces oral cancer" and assessment of oral cancer risk). We assess whether consumers respond to disease-specific modified-risk claims in a disease-specific manner. US adult current, former and never tobacco users (n=9,830) recruited from online research panels viewed advertisements stating that smokers who switch completely to Camel Snus (CS) could reduce their risk of lung cancer; a random half of respondents also viewed a claim for reduction of oral cancer risk. Respondents then rated the risk for lung cancer and oral cancer on a 1-7 scale, separately, for CS and cigarettes; the difference in the respective tobacco product types indicated perceived risk reduction. Advertisements also included four government-mandated smokeless tobacco warnings, randomly rotated. We compare those who viewed a warning about mouth cancer with those who viewed a warning about addiction. Findings indicate that the presence or absence of an oral cancer claim had no effect on perceived risk reduction for lung cancer (both groups, -1.8±0.02). With regard to warnings, respondents who viewed the mouth cancer warning reported no risk reduction for oral cancer (0.0±0.03), whereas those who viewed the addiction warning perceived lower risk for oral cancer (-0.4 ±0.03), especially if they viewed the reduced-oral-cancer claim (-0.6±0.04). These findings indicate that consumers respond to disease-specific warnings and risk-reduction claims in disease-specific ways, and suggest the value of evaluating risk perceptions in a disease-specific manner.

Determination of select leachable compounds in electronic cigarette aerosol by GCMS

WEISNER N.; ADAMS C.J.; CHOUCHANE S.
Eurofins Lancaster Laboratories PSS, Winston-Salem, NC, USA

Polymeric materials used in the manufacturing of electronic cigarette cartridges may contain compounds that have the potential to leach into the liquid or the aerosol of electronic cigarettes. A method was developed and validated for the determination of select leachable compounds such as acetone, n-hexane, methyl methacrylate, 2-ethoxy ethanol, toluene, hexamethylcyclotrisiloxane, tetradecamethylcycloheptasiloxane, butylated hydroxy toluene, and dimethyl terephthalate. The method presented here is a Gas Chromatography Mass Spectrometry (GCMS) method using Single Ion Monitoring (SIM) whereby the aerosol from the Electronic Cigarette was collected via Cambridge filter pads and impingers containing 1,2-dichloroethane which were combined after smoking for analysis. The method validation results will be presented.

Analysis of tobacco specific nitrosamines in electronic cigarette liquid by LCMSMS

SPANN N.D.; SEAMANS A.; ADAMS C.J.; CHOUCHANE S.
Eurofins Lancaster Laboratories PSS, Winston-Salem, NC, USA

A new method was needed to analyze low-level TSNA concentrations in electronic cigarette liquid. This proved difficult with the various flavored matrices, however the method was developed to accurately quantitate TSNAs as low as 50 parts per trillion for N-nitrosonornicotine (NNN), 4-(methylnitrosamino)-1-(3-pyridyl)-1-butanone (NNK), N-nitrosoanatabine (NAT) ; concentrations of N-nitrosoanabasine (NAB) are accurately quantitated at 13 parts per trillion. An Agilent 1290 Infinity II LC paired with an AB Sciex 6500+ Mass Spectrometer were used for analysis. Chromatographic separation was achieved using a Phenomenex® Gemini® 3µm C18, 110Å 150 x 2 mm column. This method has been validated and the results will be presented.

Identification of harmful and potentially harmful constituents in tobacco smoke produced from low- and high-nicotine cigarette tobacco fillers

WILSON W.B.; SANDER L.C.
National Institute of Standards and Technology, Gaithersburg, MD, USA

In 2012, the Food and Drug Administration (FDA) included 93 harmful and potentially harmful constituents (HPHCs) in the Federal Register List that are suspected to be found in tobacco products and tobacco smoke. The National Institute of Standards and Technology (NIST) has recently developed a new low-nicotine cigarette tobacco filler Standard Reference Material (SRM 3222) in collaboration with the Center for Tobacco Products (CTP) at the FDA. In July 2017, FDA announced a comprehensive plan for lowering nicotine in cigarettes to a minimally or non-addictive level. More recently, the FDA issued an advance notice of proposed rulemaking seeking public comments for consideration in developing a potential nicotine product standard to help with their plan. Within this context, SRM 3222 is currently the only low-nicotine filler reference material available. Mass fraction values have been assigned for nicotine at 0.117 ± 0.018 mg/g in SRM 3222 (as-received) based on liquid chromatography and tandem mass spectrometry and gas chromatography/mass spectrometry (GC/MS) measurements. This concentration level is approximately two orders of magnitude lower than the level of nicotine in a normal cigarette. In the current study, SRM 3222 and three high-nicotine tobacco filler materials (SRM 1082, SRM 1196, and UK 3R4F) were burned and the tobacco smoke was collected. Several HPHCs were identified in the tobacco smoke samples via GC/MS and mass fractions values were determined. Similar to the tobacco material, nicotine was determined to be two orders of magnitude lower in SRM 3222 smoke.

Effect of SBA-15 on the thermal decomposition of NNK and NNN tobacco specific nitrosamines

ASENSIO J.; MARCILLA A.; BELTRÁN M.I.
University of Alicante, San Vicente del Raspeig, Spain

Tobacco specific nitrosamines (TSNAs) are among the most carcinogenic compounds present both in tobacco and tobacco smoke, N-nitrosonornicotine (NNN), N-nitrosoanabasine (NAB), N-nitrosoanatabine (NAT) and 4-methylnitrosamino-1-3-pyridyl-1-butanone (NNK) being the most abundant. The Knowledge of their thermal behaviour and decomposition products, as well as the effect of potential catalysts for reducing their concentration is of great interest. Thus, the present study presents preliminary results on the thermal degradation of two TSNA (i.e.: NNK and NNN) under different conditions, and in the presence of SBA-15 catalysts. NNK and NNN with and without catalyst were thermally treated under two atmospheres at different temperatures using a pyrolizer (Py/EGA-GC/MS) online with a gas chromatographer coupled with mass selective detector. Flash pyrolysis was run under He atmosphere, at 300, 400 and 500 °C. Flash pyrolysis under synthetic air oxidative atmosphere was performed at 325, 375, 425 and °C. A SBA-15 mesoporous catalyst, synthetized in our laboratory, with fibre like morphology, was used. SBA-15 was characterized by N2 adsorption isotherms, scanning electron micrographs (SEM) and transmission electron micrographs (TEM) and X-ray diffraction. The presence of the catalyst under inert atmosphere promotes the decomposition of NNK and NNN and modifies the product distribution generating, in the case of NNK, myosmine, nicotyrine as the major degradation compounds, as compared with 4-Hydroxyphenyl acetonitrile in absence of the catalyst. Under oxidative atmosphere the catalyst also accelerates the decomposition of both Nitrosamines. Noting that in the case of NNN, SBA-15 favours the formation of lower molecular weight compounds.

Comparative measurements of puff aerosol size and concentration of combustion, heated tobacco, and liquid e-cigarettes by light scattering method

HOLVE D.J.(1); HOORNSTRA J.(2)
(1) Envirometrix Instruments LLC, Berkeley, CA, USA; (2) Envirometrix Instruments LLC, Sint Maartensbrug, Netherlands

In a previous presentation at the 2015 TSRC, the development and application of a new light scattering instrument based on Mie theory (Orion-A) was presented (ref 1). Further developments and improvements have been implemented on the Orion instrument, including automatic puff flow measurements based on measured pressure drop, display of complete size distribution, and size uncertainty estimation (typically 3-4%) based on three independent scattering ratio measurements. The real-time puff flow and uncertainty measurements provide improved accuracy of both size and puff mass (+/- 10%), confirmed by measurements of polystyrene latex spheres with known sizes, and by capsule mass loss measurements for liquid e-cigarettes. The poster focuses on latest results of a comparative set of measurements for a variety of nicotine delivery systems, including combustion and heated tobacco, and liquid e-cigarettes. Results show a typical mass mean aerodynamic particle sizes are in the range of 300nm for liquid e-cigarettes (LEC's), and 250nm for heated and combustion cigarettes. Puff mass measurements are generally uniform in mass delivery for LEC's over a series of puffs, while combustion and heated cigarette concentrations generally increase with puff sequence. Most LEC's show a slowly rising concentration as the heating coil is activated, although one device shows a markedly different peak concentration at the beginning of the puff. The differing size and concentration profiles indicate varying coil temperature/time profiles that can have a strong effect on the user experience and concentrations of chemical species. 1. INVESTIGATE PARTICLE SIZE DISTRIBUTIONS OF ELECTRONIC CIGARETTE AEROSOLS USING A MULTI-ANGLE LIGHT SCATTERING INSTRUMENT. Chen Song1, Don Holve2; and Steve L. Alderman1; 1R.J. Reynolds Tobacco Company, Winston Salem, NC USA, 2EnviroMetrix Instruments LLC, Berkeley, CA USA. 69th TOBACCO SCIENCE RESEARCH CONFERENCE September 20-23, 2015 Naples, Florida USA

A streamlined means of outlier detection in collaborative and proficiency studies

SLONE S.; SHELTON B.; SHEARER A.; CANETE S.
University of Kentucky, Lexington, KY, USA

Regardless of the industry, one of the first steps in analyzing data from a collaborative or a proficiency study is to detect any outliers in the data. Guidelines published by the International Organization for Standardization, ISO, enumerate several tests to consider. To identify outlying means in the laboratories participating in proficiency study, the guidelines suggest the single Grubb's test and the Mandel's h test. Similarly, the guidelines propose using the Cochran's test and the Mandel's k test to identify outlying standard deviations. While the guidelines present a partial table of significant values for these four outlier tests, a programmatic solution to find the scores and critical values leads to savings in time and effort. In his 2013 paper in AStA Advances in Statistical Analysis, Wilrich derived the formulas for the critical values for each of the aforementioned tests. A SAS® macro was written to perform these tests on outliers for data from the Center for Tobacco Reference Products (CTRP) at the University of Kentucky. The CTRP utilizes this macro for proficiency testing and for characterization/certification of constituents in tobacco reference products as part of an FDA funded Cooperative Agreement. Since the macro is written for flexibility, a user can choose the tests to be performed as well as various options for printing which include significant results for the Grubbs or Cochran test, graphs for the Mandel's h and k, with reference lines at the 95% and 99% critical values, and a summary table of results. Details on and sample runs of the macro will be presented. The macro has been updated to handle missing data.

The quantitation of volatile nitrosamines in mainstream tobacco smoke by gas chromatography-tandem mass spectrometry (GC-MS/MS) using direct sample bag collection

COWAN E.A.; WINKELJOHN W.R.; LISKO J.G.; VALENTIN-BLASINI L.
Centers for Disease Control and Prevention, Atlanta, GA, USA

Volatile nitrosamines (VNAs) are classified as group 2A and 2B carcinogens in humans by the International Agency for Research on Cancer (IARC), have been associated with increased incidences of carcinogenesis and have been linked to lipid peroxidation, oxidative stress, and insulin resistant diseases. Potential sources of VNA exposure include contaminated water, cosmetics, food products, as well as tobacco products. Literature has shown that smoking cigarettes or using smokeless tobacco products may cause higher exposure to VNAs compared to other common sources. As a result, the U.S. Food and Drug Administration listed multiple VNAs as Harmful and Potentially Harmful Constituents (HPHC) in tobacco products and tobacco smoke. Previously reported methods of VNA quantitation used complex protocols that required large sample sets of cigarettes, tobacco smoke collection using impingers, and using treated smoking pads, which causes decreased sample throughput. Additionally, previous methods utilize a non-specific method of detection (gas chromatography-thermal energy analyzer), which may affect VNA quantitation. We report the development and validation of a specific and sensitive method for the detection of 8 VNAs (N-nitrosodimethylamine (NDMA),N-nitrosodiethylamine (NDEA), N-Nitrosomethylethylamine (NMEA), N-nitrosodipropylamine (NDPA), N-nitrosomorpholine (NMOR), N-nitrosopyrrolidine (NPYR), N-nitrosopiperidine (NPIP), and N-nitrosodibutylamine (NDBA)) in mainstream tobacco smoke using Tedlar® bag smoke collection, one-step SPE sample cleanup, and GC-MS/MS for quantitation. Method validation was completed for both ISO 3308 (IS) and Canadian Intense (CI) smoking regimens and the method was applied to 20 commercial brands of cigarettes and little cigars. Method precision and accuracy was 5-28% RSD and 84-111%, respectively, and the limit of detection (e.g., NDMA, 0.931ng/cig (ISO) and 0.693 (CI); NMEA 0.0267 ng/cig (IS) and 0.0279 (CI)) was improved for most analytes compared to previously reported methods.

Cumulative probabilistic risk assessment to characterize risk of combustible tobacco products

KOSARAJU K.; AYALA-FIERRO F.; ROBINSON E.; TURNER A.
ITG Brands LLC, Greensboro, NC, USA

Quantitative risk assessment (QRA) is an established deterministic approach which uses point estimates that represent different points in a statistical distribution, the median (often referred as the likeliest value) or more conservative point estimates such as the 95 or 97.5% percentile values, and does not account for the distributions applicable to smoking behavior and exposure variables used in risk assessment models. Probabilistic risk assessment (PRA) employs a probability based approach to address the contribution of variability and uncertainty related to assumptions used to compute the overall risk, Incremental Lifetime Cancer Risk (ILCR) or Hazard Quotient (HQ), for cancer and non-cancer endpoints, respectively. Cumulative or aggregate risk assessment is employed during risk assessment of chemical mixtures. The objective of this presentation is to present a HPHC20 based cumulative PRA approach to compare the aggregate risk of two combustible tobacco products and evaluate the approach in comparison to risk assessment of individual analytes.

Survey of nicotine, NNK, NNN, pH and physical properties of little cigars in the US market place

LIU Youbang(1); HASSINK M.D.(1); TAYLOR K.M.(1); WATSON C.(2); CHAFIN D.(2); LAWLER T.(2); VALENTIN-BLASINI L.(2); KIMBRALL B.(2)
(1) U.S. Food and Drug Administration, Silver Spring, MD, USA; (2) Centers for Disease Control and Prevention, Atlanta, GA, USA

In May 2016, the jurisdiction of the Federal Food Drug and Cosmetic Act was extended to include all tobacco products, including cigars. Because limited information exists on the chemistry of cigar filler and smoke, 60 brands of popular little cigars sold in the US were characterized for physical properties (i.e., tobacco weight, rod length, and rod diameter), tobacco filler nicotine, NNK, NNN, and pH, using validated in-house methods. Preliminary analysis shows mean tobacco weight at 1085 (range 767 – 1367) mg; mean rod length at 97.4 (range 73.9 – 100) mm; mean rod diameter at 7.8 (range 6.8 - 8.4) mm; mean concentrations for: 1.) nicotine at 12.7 (range 7.9 - 21.0) mg/g tobacco; 2.) NNK at 1118 (range 26 - 2950) ng/g tobacco; 3.) NNN at 4714 (range 1440 - 12100) ng/g tobacco; and mean pH at 6.2 (range 5.4 - 6.6). Using tobacco pH, the mean free nicotine concentration was determined to be 0.21 (range 0.05 - 0.33) mg/g tobacco, with the mean percent free nicotine at 1.7% (range 0.2% - 3.8%). Unlike other types of cigars, little cigars are similar to cigarettes with respect to size and components. A preliminary comparison of the 60 little cigars with 50 popular cigarettes shows differences between these product categories. The mean total nicotine concentration of little cigars is lower than that of cigarettes (12.7 vs. 19.2 mg/g tobacco), while the mean NNK and NNN concentrations of little cigars are higher (1118 vs 523 and 4714 vs 1901 ng/g tobacco, respectively). The mean pH of the little cigars is greater than that for cigarettes (pH 6.2 versus pH 5.5), causing them to have a higher mean free nicotine than the cigarettes (0.21 vs. 0.05 mg/g tobacco; 1.7% vs. 0.3% for the percent free nicotine). The data suggest that despite their similarities to cigarettes, little cigars may deliver more free nicotine and expose users to more harmful chemicals.

Assessment of in vitro toxicity/genotoxicity of ends and combustible products

JORDAN K.(1); BOMBICK B.(1); COOPER B.(1); LEVERETTE R.(1); SHUTSKY T.(1); WEST C.(1); STRASSER J.(1); ROUND E.(1); MARANO K.M.(1); MOMIN R.(2); TRIVEDI A.(2); WAGSTAFF W.(2); BRUCE S.(3); DAKOULAS E.(3); ROY S.(3)
(1) RAI Services Company, Winston-Salem, NC, USA; (2) Labstat International ULC, Kitchener, ON, Canada; (3) BioReliance Corp, Rockville, MD, USA

In May 2016, the U.S. Food and Drug Administration Center for Tobacco Products (CTP) issued draft guidance for Premarket Tobacco Product Applications (PMTA) for Electronic Nicotine Delivery Systems (ENDS), which recommends inclusion of in vitro toxicology testing (e.g., genotoxicity and cytotoxicity studies). Specific to genotoxic potential, the draft guidance suggests using assays/methods outlined in International Conference on Harmonization (ICH) S2(R1) and Organization for Economic Cooperation and Development (OECD) guidelines. The draft guidance also recommends testing multiple test article concentrations in each assay, and using comparator products for hazard identification. We present an example of in vitro toxicology testing to support an ENDS PMTA. Bacterial mutagenicity (Ames test), in vitro micronucleus and cytotoxicity assays were conducted on ENDS e-liquids, ENDS product aerosols and U.S. market comparators in general accordance with applicable guidelines (ICH S2(R1); OECD 487 and 471; NIH Publication 07-4519 [2006] and Health Canada Official methods (T-501/T-502/T-503). Aerosol total particulate matter (TPM) and gas vapor phase (GVP) from ENDS products, and a combustible comparator were generated under intense puffing/smoking regimens for testing. Neat e-liquid samples were diluted and applied to the test systems. Test samples were evaluated up to concentrations resulting in toxicity or up to limits inherent in the assays. The ENDS e-liquids, TPM and GVP were either negative or demonstrated limited response at the highest concentrations tested in each of the assays, indicating a lack of toxicity and genotoxicity of the ENDS products under the experimental conditions tested. In contrast, with the exception of GVP in the Ames assay, the combustible market cigarette comparator was consistently positive, confirming that the assays conducted are sensitive to potential toxicants/genotoxicants found in combustible tobacco product emissions.

Evaluation of filter density vs filter pressure drop for characterizing cellulose acetate filters

NORFLEET K.R.; ROBERTSON R.
Celanese Corporation, Narrows, VA, USA

The consideration of filter density as a filter design parameter has increased in recent years. Traditional filter design is based on pressure drop or resistance to draw to achieve specified smoke filtration levels (example NFDPM). Pressure drop is related to the acetate fiber surface area which filters main stream smoke. Filter density does not necessarily correlate with fiber surface area. For example, changing the filter length will not change the density but has a significant impact on filter pressure drop (fiber surface area). Additionally, under or over processed the tow has an impact on filter density and filter pressure drop. Incorrectly processed tow is easily detected by filter pressure but filter density is an unproven parameter. This poster will illustrate the impact of the filter physical dimensions, plug wrap paper thickness, tow item, and tow processing on density and pressure drop. Under certain conditions filter density may be an appropriate way to model filter performance but may not be suitable under all conditions.

Abuse liability assessment of three non-menthol ends products relative to combustible cigarettes and nicotine gum

CAMPBELL C.; JIN Tao; WHELEN M.J.; EVANS E.H.; BAXTER S.
RAI Services Company, Winston-Salem, NC, USA

An abuse liability (AL) assessment of tobacco products, including electronic nicotine delivery systems (ENDS), may be based on a pharmaceutical model that includes subjective measures and pharmacokinetic effects after short-term product use. This study assessed the AL of three tobacco-flavored Vuse ENDS products (Vuse Vibe [tank, 3% nicotine], Vuse Solo [cigalike, 4.8% nicotine] and Vuse Ciro [cigalike, 1.5% nicotine]) in comparison to usual brand (UB) cigarettes and nicotine gum in a population of smokers with limited ENDS experience. During the 11-day single-site, randomized, crossover confinement study, subjects used each product for 1.5 days of trial use, followed by overnight abstinence and a 4-hour Test Session. During each Test Session, subjects used a single product for a specified period of time and were assessed for subjective measures including product liking over time, maximum product liking, overall intent to use again, and positive and negative product effects (PE). Nicotine uptake measures Cmax, Tmax, and AUCnic0-240 were also assessed. Product liking scores were statistically lower for the Vuse products than for UB, but higher than for nicotine gum. Positive PE for Vuse scored between UB and gum; negative PE scores were similar to UB and generally lower than those for gum. When compared to UB, nicotine uptake measures were significantly lower for Vuse products; however, nicotine uptake for Vuse was similar to or slightly lower than for nicotine gum. Similar to previous findings for Vuse Solo with varying nicotine content, Vuse product use resulted in subjective effects and nicotine uptake measurements between those of UB and nicotine gum, suggesting that the abuse liability of these products is likely lower than combustibles and somewhat higher than nicotine gum.

Neutral red uptake (NRU) cytotoxicity analysis of aerosol generated from a temperature-regulated nicotine-salt based ends product

MISRA M.(1); GILLMAN I.G.; DESAI P.
(1) JUUL Labs, San Francisco, CA, USA; (2) Enthalpy Analytical, Inc., Richmond, VA, USA

JUUL ENDS is a temperature-regulated product which was designed to minimize the generation of combustion related degradation byproducts in the aerosol. The overall purpose of this study was to test the cytotoxicity of the CHO-K1 cells in response to 24h exposure to various concentrations of ENDS aerosol using the NRU in vitro cytotoxicity assay using CHO-K1 cells according to OECD guidelines (TG 129). Methods: CH0-K1 cells were treated with ENDS aerosol, positive control, or vehicle control. JUUL ENDS aerosol was tested for three different flavor profiles. The cells were exposed to the different test articles; ENDS aerosol at the concentration of (0- 300 µg/mL) and compared with the mainstream smoke component (TPM) from 3R4F, at dose range of (0-150 µg/mL). Additionally, the cells were exposed to the positive control, SLS at concentrations 40 and 80 µg/mL. Results: Amongst all of the different samples tested, no aerosol mediated significant toxicity was observed at any of the concentrations tested for the test articles, Classic Menthol, Cool Mint, Cool Cucumber, as well as for PG/VG/Benzoic Acid/Nicotine (Carrier Control). At the doses tested, EC50 for the e-cig aerosol and carrier control aerosol could not be calculated because cell viability was greater than 70% at all concentrations tested. Additionally, TPM from the 3R4F cigarette which showed the expected toxicity with a calculated EC50 of 59.46 µg/mL with r2 = 0.98. In summary, under the experimental conditions and based on the criteria for Evaluation of Cytotoxic Response (ISO 10993-5), the test articles Classic Menthol, Cool Mint, Cool Cucumber, as well as for PG/VG/Benzoic Acid/Nicotine (Carrier Control) are considered non-cytotoxic.

Challenges and opportunities in cigar science

SMITH J.H.(1); AUBUCHON S.M.(1); WAGNER K.A.(1); McKINNEY W.J.(1); GILLMAN I.G.(2); STEVENS R.(3)
(1) Altria Client Services LLC, Richmond, VA, USA; (2) Enthalpy Analytical, Henrico, VA, USA; (3) ITG Brands LLC, Greensboro, NC, USA

An independent Cigar Working Group was formed by interested cigar stakeholders (e.g. manufacturers and testing laboratories). The goals of the Working Group are to explore opportunities for engagement with the Food and Drug Administration ("FDA") on issues relating to the manufacturing, analysis and testing of cigars. Recent deeming by the FDA places the burden on cigar manufacturers to obtain authorization for new products not on the market as of February 15, 2007 through one of three regulatory pathways. However, scientific research necessary to support regulatory submissions for cigars is limited and some of what exists, due to a lack of standardization in the field, is lacking in rigor. Though cigars and cigarettes are both combustible tobacco products, cigars, unlike cigarettes, have not been the subject of widespread scientific study. A Cigar Science sub-group was formed within the Cigar Working Group to explore the challenges and opportunities within the cigar science space. The group has identified the following as areas where research and development is warranted: 1. Quantity and breadth of scientific literature on cigars 2. Cigar use patterns and smoking topography 3. Cigar reference products 4. Standardized, validated analytical methods for chemical constituents 5. Lack of a common set of defined terms or terminology The outcome of this collaboration will ensure that: a) future research on cigars is standardized across laboratories, b) cigar terminology will be consistent and c) future cigar research is robust and reproducible. These key factors will ensure that questions of public health are appropriately answered.

Effect of changes in cigarette input parameters and distributions on probabilistic risk assessment

ROBINSON E.; AYALA-FIERRO F.; KOSARAJU K.; TURNER A.
ITG Brands LLC, Greensboro, NC, USA

Probabilistic Risk Assessment (PRA) uses a probability based approach to examine the contribution of variability and uncertainty in assumptions used to compute the risk for cancer and non-cancer endpoints. PRA is performed by introducing distribution variability associated with constituent yields and exposure parameters (Cigarettes per Day, CpD; Exposure Duration, ED; Averaging Time, AT; Body Weight, BW; and Daily Inhalation Rate, DIR), which are often "worst-case" estimates. As cigarette use and population demographics have changed over time, the potential input parameters and their distributions have also changed. Cigarette use (CPD) has been declining for many years while life expectancy, which influences ED and AT, and BW, which influences DIR, have increased. This presentation aims to examine the impact of changes in these input parameters and their distributions on risk assessment for cigarettes using a PRA approach.

Estimation of corrosion of metal wires in e-liquids

THORN E.K.; RUSYNIAK M.J.
Altria Client Services, LLC, Richmond, VA, USA

Metal content in e-cigarette liquids and aerosols has been drawing the attention and concern of the public health community and tobacco industry. One possible source of metal contamination could be due to the corrosion or degradation of the metal heater coils when in contact with e-liquids during either storage or vaping. Metal coils can range in composition, but are commonly made from alloys containing nickel, chromium, and iron. An electrochemical method using linear polarization to evaluate corrosion propensity of metal alloys in e-liquid formulations is presented. Additional factors of alloy composition and formulation pH were also investigated. The method presented is suitable for rapidly screening the compatibility of metals and e-liquids.

Nicotine uptake in naïve, short-term and experienced ends users compared to nicotine uptake in smokers

HONG J.(KS); JONES B.A.; BAXTER-WRIGHT S.; JIN Tao; CAMPBELL L.
RAI Services Company, Winston-Salem, NC, USA

Consensus is growing among researchers and public health organizations, including the National Academies of Science, Engineering and Medicine, and American Cancer Society, on harm reduction from the use of electronic nicotine delivery systems (ENDS) as compared to cigarette smoking. The US FDA Center for Tobacco Products recommends in draft guidance that human studies to support Premarket Tobacco Product Applications for ENDS include pharmacokinetic (PK) data examining the exposure to nicotine during ENDS use. Some studies indicate that experienced ENDS users reach levels of nicotine more similar to users of combustible cigarettes, albeit generally lower. In this study, selected nicotine PK measures (AUCnic 0-360, Cmax, Tmax) were investigated to assess nicotine exposure in naïve, short-term and experienced ENDS users over 6 hours with respect to the start of a 10-minute ad libitum product use period. The ENDS products were Vuse Solo Original and Vuse Solo Menthol. Data was compared to historical PK data from non-menthol and menthol cigarette smokers. Nicotine uptake (AUCnic 0-360 and Cmax) was significantly higher in both non-menthol and menthol cigarette smokers compared with Vuse Original and Vuse Menthol experienced users, respectively. Vuse Original and Vuse Menthol naïve users also had significantly lower nicotine uptake compared with experienced Vuse users. PK parameters in experienced and short-term Vuse Original and Vuse Menthol users were similar, with the exception of Vuse Original AUCnic 0-360, which was significantly higher in experienced users compared to short-term users. Tmax occurred significantly faster in the non-menthol and menthol cigarette smokers compared with the experienced Vuse Original and Vuse Menthol users. However, differences in Tmax between the experienced versus short-term or naïve Vuse users were not statistically different.

Real-time authentication of food, beverages and botanicals using DART-QDa LiveID analysis

BRITTON E.(1); ORGANTINI K.(1); MERUVA N.(1); STEAD S.(2)
(1) Waters Corporation, Milford, MA, USA; (2) Waters Corporation, Wilmslow, UK

This study investigates the potential for the implementation of DART-QDa with LiveID as solution for the rapid authentication of high value food, beverage and botanical ingredients and products with minimal sample preparation. Food, beverages and botanicals of higher commercial value are frequently subject to fraudulent practice. Analytical methods are required for process control, quality assurance and flavor/formulation monitoring. Analysis of the characteristic profiles of such high value products can be conducted using extraction followed by chromatography for authenticity purposes. Such methods are time-consuming and there is a need for rapid, qualitative testing to keep pace with industry requirements. DART produces relatively simple mass spectra characterized by M+. Or [M+H]+ in positive mode and M-. , [M+H]- in the negative mode. The mass spectral features are used to generate a multivariate model using a combination of principal component analysis (PCA) for data dimension reduction and linear discriminant analysis (LDA) for class discrimination to distinguish between samples of different processing and production types.

Compartmental analysis of inorganic and organic constituents and the dithiothreitol-measured oxidative potential of waterpipe tobacco smoke

HAUSER C.D.; CONQUEST P.; JENSEN A.; MERRITT A.; PELAEZ O.; PENG C.; POHL N.; TAYLOE E.; WOOD S.; BERND K.
Davidson College, Davidson, NC, USA

With Hookah's growing acceptance in the U.S., the general age of hookah smokers is getting younger – 18% of high school seniors claim that they have used waterpipe tobacco smoking (WTS). As a result of its pleasant manufactured flavor, social acceptance and prevalence of Hookah cafes, influence of social media, and lack of restriction policies, Hookah's negative effects are often ignored. Hookah can cause long term and acute damages on the human cardiovascular system and respiratory system, and can possibly lead to increased blood pressure, lung inflammation, Ischaemic heart disease, chronic bronchitis, lung cancer and many other diseases. Research in our laboratories to date have focused on the potential physical toxicity associated with particles in waterpipe tobacco smoke. However, when the physical profile of a WTS sample fails to explain its effects on cell health, the chemical characteristics of the sample may be responsible for its cytotoxicity. Chemical characterization of WTS to date has been restricted to targeted analyses of toxicants also present in cigarette smoke. These targeted studies are a reasonable preliminary step, yet they fail to account for toxicants unique to WTS. Waterpipes employ a different smoke-producing mechanism and contain different additives (such as flavorings and syrups) than cigarettes, thus producing different chemical profiles. Here we present the results of non target analyses of semivolatile organic compounds using LCMS methods as well as a compartmental analysis of metals in the waterpipe system using FAAS to evaluate the potential chemical toxicity of WTS. We have coupled this chemical analysis with the dithiothreitol (DTT)-measured oxidative potential of the particulate component of WTS.

Evaluation of total particulate matter and whole aerosol from combustible and next generation tobacco products (NGPs) in in vitro lung models

FIELDS W.; KEYSER B.; BOMBICK B.
RAI Services Company, Winston-Salem, NC, USA

Strategies for 'Toxicity Testing in the 21st Century' include the use of human cellular systems to evaluate the health risks of potentially toxic chemicals. Therefore, the goals of this study were to assess toxicological responses from combustible cigarettes and electronic nicotine delivery systems (ENDS) in 2D and 3D normal human bronchial epithelial (NHBE) cells. Total particulate matter (TPM) and whole smoke/aerosol (WS/WA) were generated from Kentucky reference (3R4F) cigarettes and two market ENDS under HCI or CRM81 regimes, respectively. NHBE-2D cells were subsequently exposed to 1–150 ug/ml TPM or ENDS TPM (eTPM) in submerged cultures, and EpiAirwayTM (NHBE-3D) cells were exposed to WS/WA at air-liquid interface. Cytotoxicity was evaluated with the neutral red uptake (NRU) assay in NHBE-2D cells and with the 3-(4,5-dimethylthiazol-2-yl)-2,5-diphenyltetrazolium (MTT) assay in NHBE-3D cells. Gene expression was also assessed following WS/WA exposures using quantitative polymerase chain reaction. 3R4F TPM yielded a dose-dependent decrease in NRU in NHBE-2D cells with an IC50 of 51.8+7.3 ug/ml. However, an IC50 was not achieved for the eTPMs. Similarly, viability of NHBE-3D cells was reduced dose-dependently by 3R4F WS with an IC50 value of 21.8+2.0 puffs, while viability remained >95% following WA exposure from 1-180 puffs of ENDS. Expression-profiles of genes associated with oxidative stress, inflammatory response, chemical activation and detoxification (NQO-1, SPP1, CYP1A1 and HMOX-1) were increased 2- to 120-fold by 3R4F WS compared to the air control. In contrast, the ENDS WA yielded < 1.2-fold for the same genes with the exception of CYP1A1, where a 2.6-fold increase was observed. These data suggest that assessments in in vitro lung models may support understanding of the differential toxicological impact of combustible cigarettes and NGPs.

Innovative filter design for tobacco heated products (THPs)

PERIMAN A.; TREADAWAY A.; DILLIE B.; GERRINGER E.; BOLTON G.
Essentra Filter Products, Greensboro, NC, USA

Tobacco Heated Products (THPs) are quickly growing as part of the tobacco market. These products fall into the Heat not Burn category, which requires some differences in the filters utilized. THPs do not require much actual filtration, since the tobacco is not combusted like in a traditional cigarette. The filters are mainly present to help cool the vapour before it reaches the consumer's mouth. With this different requirement, many producers are looking for low pressure drop options, such as hollow acetate tubes. With this being said, Innovation is not limited in this sector, with many customers exploring options with flavors, capsules, shapes, alternate filtering media, and many others.

Observation of wicking behavior of an ends device using weight-time measurements

PLATT S.P.; THORN E.K.; RAGLAND R.B.; RUSYNIAK M.J.
Altria Client Services LLC, Richmond, VA, USA

The rate of liquid uptake (wicking) of an e-vapor liquid is an important factor for the performance of an electronic nicotine delivery system (ENDS) device. A common method for quantifying the wicking rate uses Lucas-Washburn theory. Here the time required for a liquid to travel a certain height in the material due to capillary action is determined. An alternative to this traditional height-time approach to measuring the wicking rate uses the mass uptake of the liquid by the material over time. Weight-time measurements consider the rate of liquid traveling throughout the entire wick whereas height-time measurements only consider a single observable edge. A modified Lucas-Washburn equation serves as a model equation for these weight-time measurements. This modified equation gives a capillary constant, an effective composite rate constant for the entire wicking process that includes both liquid parameters and wicking dimensions. The mass uptake of a variety of e-liquids by a wick made of fiberglass strands offers a test case to apply this method. A tensiometer serving as a hanging balance monitored the mass uptake of an e-liquid into an ENDS device wick over the time interval required for saturation. The calculated capillary constants from these experiments correlate with the liquid parameters of the modified Lucas-Washburn equation: viscosity, surface tension, and liquid density. The results show that weight-time measurements offer a direct and reproducible method to understanding the wicking process observed in ENDS devices.

Non-targeted analysis using gas chromatography mass spectrometry to evaluate stability of e-vapor products

CROSSWHITE M.; SHAH N.; NOE M.; SUAREZ C.; MILLER J.H. IV; PITHAWALLA Y.B.; GARDNER W.P.
Altria Client Services LLC, Richmond, VA, USA

In the newly regulated landscape of tobacco products there is an increasing need to fully characterize the chemical composition of aerosol in new products. Although aerosol from e-vapor products are considerably less complex than aerosols from Heat-Not-Burn (HNB) or mainstream smoke from cigarettes; there are still challenges that arise from the chemical composition of the matrix, variety of flavors used, and the potential for chemical interactions to occur during storage. There is additional complexity associated with efficiently collecting both volatile and semi-volatile compounds delivered in the aerosol from e-vapor products. We have developed an automated workflow for data analysis that includes mass spectral deconvolution, peak detection, library searching and reporting. Aerosol samples were collected using a 55 mm Cambridge filter pad (CFP) with a trailing impinger containing 10 mL of ethanol chilled to -70 °C, to ensure both volatile and semi volatile compounds are captured. Samples were analyzed on an Agilent GC/MS system (7890B with 5977A) using a Restek Stabilwax® GC column (30 meter x 0.25mm ID x 0.25µm film) with an infused 5 meter integra guard column. Our workflow includes both Agilent MassHunter Unknowns Analysis and AMDIS (Automated Mass Spectral Deconvolution and Identification System) software for identification of extraneous peaks. Identification of compounds not present in mass spectral libraries includes secondary analysis using high resolution mass spectrometry on a GC-OrbitrapTM with EI and CI ionization modes, allowing for the identification of molecular formulas within 5 ppm of mass accuracy. Compound identification was confirmed through the use of reference standards.

Risk perception of e-vapor and the impact of promotional materials

ZDINAK J.P.; GOGOVA M.; LEIGHTY J.; PLUNKETT S.
Altria Client Services LLC, Richmond, VA, USA

FDA's new comprehensive tobacco and nicotine strategy acknowledges the risk continuum and fosters innovation in noncombustible lower risk tobacco alternatives. Risk perceptions play a role in trial and/or adoption of potential reduced risk tobacco products, such as e-vapor. Therefore, it is important for research to assess risk perceptions of tobacco products across a continuum of risk. In the present research, we examined the effect of MarkTen® promotional materials on adult tobacco users (n = 1,688) and non-users' (n = 1,252) perceptions of risk. Participants were assigned to one of two conditions, each with a different level of exposure to promotional materials. The first condition exposed participants to a full assortment of promotional materials, whereas the second condition exposed participants to pack shots. Participants completed pre- and post-test surveys measuring relative risk perceptions. Roughly 40% of adult smoker subgroups and 50% of non-users indicated that MarkTen® e-vapor/e-cigarette products are equally or more risky than cigarettes. Pre-post changes were minimal and no major differences appear between the two conditions. These results support a growing body of evidence showing misperceptions of risk among adult smokers. For tobacco harm reduction to be successful, it is critical that adult smokers have access to truthful and accurate information about relative risk to make an informed choice.

Distinct gene expression changes in peripheral blood mononuclear cells exposed to combustible versus non-combustible tobacco product preparations

TRAN N.Q.(1); ARIMILLI S.(2); MADAHIAN B.(3); HOMAYOUNI R.(3); PRASAD G.L.(1)
(1) RAI Services, Winston-Salem, NC, USA; (2) Wake Forest University, Winston-Salem, NC, USA; (3) Quire Inc., Memphis, TN, USA

Cigarette smoking exerts diverse physiological effects including immune suppression. Existing epidemiological data show that consumption of smokeless tobacco products, such as moist snuff, is less harmful relative to smoking. We previously demonstrated that smokers exhibit distinct peripheral blood mononuclear cell (PBMC) gene expression patterns relative to moist snuff and non-tobacco consumers. To better characterize the biological effects associated with different tobacco products, a genome-wide gene expression study was performed using a PBMC in vitro model. We investigated global gene expression changes in PBMCs exposed to different nicotine equivalent doses (low, medium, and high) of whole smoke-conditioned media (WS-CM) or one nicotine equivalent dose of smokeless tobacco extract (STE) prepared from reference tobacco products (3R4F and 2S3, respectively). Pairwise statistical analysis did not identify significant expression changes by low WS-CM compared to media control. However, transcripts were significantly affected by medium WS-CM (479 transcripts), high WS-CM (2,703), and STE (2,156) treatments. The overlap between medium WS-CM and STE, and high WS-CM and STE, was minimal (34 and 65 transcripts, respectively), suggesting distinct effects of WS-CM and STE treatments. Hierarchical clustering showed that gene expression profiles under STE clustered with those under medium WS-CM, while the highest dose of WS-CM clustered distinctly from the medium WS-CM and STE conditions. Functional analysis revealed that WS-CM exposures, but not STE, uniquely affected genes involved in immune cell development and inflammatory response. Cascades of upstream regulators (e.g., TNF, IL1β, NFκB) were identified for the observed gene expression changes and generally suppressed by WS-CM, but not STE. Collectively, these findings suggest that combustible and non-combustible tobacco products produce distinct biological effects, which could explain the observed chronic immune suppression in smokers.

A science and evidence based nicotine tobacco product standard

McKINNEY W.J.; BRANNAN A.; SMITH D.
Altria Client Services LLC, Richmond, VA, USA

Unprecedented in its complexity and scale, a very low nicotine content (VLNC) standard would fundamentally change the design and function of cigarettes, modify the choices and behaviors of more than 38 million U.S. adult smokers, and transform the market. The reduced-nicotine product standard envisioned in the Advance Notice of Proposed Rulemaking (ANPRM) will require significant science and evidence to support FDA's determination that such a standard would be appropriate for the protection of public health. FDA has stated that it will use the "best available science" to pursue any nicotine product standard. At this time however, there are several fundamental problems. The scientific data on VLNC cigarettes is limited and much of the data that FDA references in the ANPRM is not publicly available. This presentation will focus on the need for science that can be accessed, validated, authenticated, reproduced and challenged in the development of a nicotine standard.

Tobacco industry efforts to reduce nicotine

DELOACH L.
Altria Client Services LLC, Richmond, VA, USA

In March 2018, the United States Food and Drug Administration (FDA) issued an Advanced Notice of Proposed Rule Making (ANPRM) seeking comments associated with reducing the nicotine content of combustible tobacco product filler to a level that is minimally or non-addictive. Cigarettes comprise the largest majority of combustible products sold in the U.S. and most consist of an 'American blend' comprised of Flue Cured, Burley, and Oriental tobaccos. Filler nicotine levels for these cigarettes typically range from ~14 to ~27 mg/g. FDA has proposed very low nicotine (VLN) levels of 0.3, 0.4, or 0.5 mg/g in filler; nearly a 98% reduction from current levels. FDA has indicated that such levels are technically achievable by employing and combining a number of potential technical solutions including, (1) agronomic practices such as "controlled growing conditions, fertilization, [and] harvest" and "cross-breeding plants," (2) tobacco blending, (3)"nicotine extraction," and (4) "genetic engineering." 83 Fed. Reg. at 11831. Numerous research studies throughout the last 60 years have concluded that although some level of nicotine reduction is achievable, reductions to levels suggested by the FDA are not technically achievable at a commercial scale. In addition, use of 'very low nicotine' tobaccos in cigarettes results in poor sensory characteristics and subsequently, rejection by adult smokers. Additionally, these studies also concluded that the potential for unintended consequences exists. This presentation will discuss previous efforts by the tobacco industry to develop tobacco plants and cigarettes having reduced nicotine levels and will address the current state of technology associated with such efforts.

Potential adverse effects of ANPRM's on small tobacco product manufacturers

LAUTERBACH J.H.
Lauterbach & Associates, LLC, Macon, GA, USA

A small tobacco product manufacturer (STPM) is defined in the Family Smoking Prevention and Tobacco Control Act (Public Law 111-31) as a tobacco product manufacturer that employs fewer than 350 employees. This category includes numerous manufacturers of cigarettes, filtered cigars, e-vapor products, and other tobacco products. Products from such manufacturers are only a small fraction of those produced by the major manufacturers and often have only regional distribution as opposed to national distribution. In general, such manufacturers do not have the scientific support staffs and laboratories typical of the major manufacturers. Thus, any change in product design, composition, and/or testing required by regulatory bodies usually means use of consultants and commercial testing laboratories. These can be major expenses for such organizations. Changes in tobacco nicotine content and/or restrictions on flavors as suggested in the ANPRM's could mean the end of such companies either through rejection of their products by consumers due to poor sensory attributes, competition from contraband products with acceptable sensory properties, and/or increased testing requirements to show that products meet new regulations, and except for mandated changes, meet other aspects of the Substantial Equivalence regulations. Moreover, the calls for removing flavors from e-vapor products misses the point that unlike tobacco products, there is no flavor (tobacco or otherwise) in unflavored e-vapor products. Furthermore, removing flavors from manufactured cigars would break a long-standing practice of adding flavors to such products. Specific examples of the problems that would be faced by STPM's and examples of alternate approaches will be given in the presentation.

Production of very low nicotine burley tobacco: short term feasibility from an agronomy viewpoint

JACK A.M.; FISHER C.R.
University of Kentucky, Lexington, KY, USA

The FDA has issued an ANPRM to obtain information for developing a product standard that would lower nicotine to non-addictive levels, and requested comments on a limit of around 0.3-0.5 mg/g. We do not believe that this limit is technically feasible on a nationwide scale in the short term with the varieties currently available. Nicotine levels are extremely variable between crops. We examined data collected in the Minimum Standards Program. Nicotine levels were mostly 35 – 55 mg/g, but they ranged from 18 mg/g to 72 mg/g. These data are weighted means for the whole plant; the leaf grade, which comprises most of the weight, is at least 12-15% higher. LA (low alkaloid) mutants reduce nicotine to about 10% of the wild type: 3.5 – 5.5 mg/g in the average crop, which is still tenfold higher than the limit in question. Agronomic practices can reduce nicotine further, but results are notoriously variable. Data we have collected over the years show: a) not topping reduced nicotine to 55-60% of topped tobacco b) close spacing reduced it to 65-95% of standard spaced tobacco c) drastically reducing nitrogen fertilizer reduced it to 40-94% of standard fertilized tobacco d) early harvesting reduced it to 64-78% of standard harvested tobacco. We are currently testing the combined effect of all these factors. We do not yet have any data, but it is unlikely that it will reduce nicotine to below 10% of standard-grown tobacco, and even more unlikely that it will do so consistently. Several groups are working on molecular nicotine reduction. It is possible that they will achieve the 0.3-0.5 mg/g nicotine levels, but it could be years before such varieties are available to growers on a commercial scale, if the industry maintains its current testing procedures.

Agricultural practices and environmental factors that affect alkaloid accumulation in tobacco

RIDDICK M.G.
R.J. Reynolds Tobacco Company, Winston-Salem, NC, USA

The U.S. Food and Drug Administration (FDA) recently issued an advance notice of proposed rulemaking (ANPRM) seeking input into the potential development of a product standard to lower nicotine in combustible cigarettes to minimally or non-addictive levels. This presentation reviews agricultural practices and environmental factors and their effect on accumulation of nicotine, the major alkaloid in tobacco. Nicotine is synthesized in the roots of the tobacco plant and is subsequently transported to leaves and stems where it accumulates. Any agricultural practice that increases the size of the root system in relationship to the size of the plant will, in general, increase the amount of nicotine accumulated in the tobacco leaves. Conversely, any agricultural practice that increases the size of the plant in relationship to the root system will, in general, decrease the amount of nicotine accumulated in the tobacco plant leaves. Other agricultural practices that have an effect on nicotine accumulation are: nitrogen rates, plant populations, topping (expressed as number of leaves per plant), and sucker control. Other factors contributing to the amount of nicotine accumulated in tobacco are environmental, among which water (in the form of rainfall), is the most significant. During an extremely dry year, tobacco will be higher in nicotine and lower in sugar content. In an extremely wet year, tobacco will be lower in nicotine and higher in sugar content. Other environmental factors include: soil type, root injury, and day length. This presentation will provide an overview of the various agricultural and environmental factors that have the greatest impact on nicotine accumulation, and will provide insight into some of the challenges that the tobacco industry and growers would face in the event of a very low nicotine product standard.

FDA advance notice of proposed rulemaking for nicotine level of combusted cigarettes – feasibility for compliance via tobacco plant genetics?

HIATT E.
R.J. Reynolds Tobacco Company, Winston-Salem, NC, USA

On March 16, 2018, the U.S. Food and Drug Administration (FDA) published an advance notice of proposed rulemaking (ANPRM) entitled "Tobacco Product Standard for Nicotine Level of Combusted Cigarettes." The publication of an ANPRM solicits information and public comments from stakeholders –industry, public health advocates, etc. – regarding a particular issue. After all of the requested information has been submitted, the FDA must consider the information in fashioning a proposed rule, or it may decline to issue a proposed rule. This presentation centers on one aspect of the referenced ANPRM – establishing a nicotine level standard in combustible cigarettes. Specific potential nicotine levels mentioned in the ANPRM are: 0.3, 0.4, and 0.5 mg nicotine/g of tobacco filler. Can levels this low be achieved in the tobacco plant and maintain tobacco leaf yield and quality? Low nicotine tobacco lines containing the nic1/nic2 loci have been available for several years, but agronomic data from large field trials is scarce and nic1/nic2 will not achieve the lowest levels mentioned in the ANPRM. Molecular biology techniques have been employed to elucidate information for many of the major nicotine biosynthesis pathway genes as well as transporters, transcription factors and non-coding RNAs. By providing an overview of these various genetic aspects and their feasibility, insight into some of the challenges facing the tobacco industry will be presented. Challenges are to not only comply with the potential nicotine level standard, but maintain both flavor characteristics acceptable to adult tobacco consumers as well as crop yields and tobacco leaf quality tobacco producers require.

Modification of standardized methods for the measurement of nicotine in very low nicotine content cigarettes

BROWN A.P.(1); MELVIN M.S.(1); PITHAWALLA Y.B.(1); SMITH J.H.(1); WAGNER K.A.(1); LIU Hui(2); MASON K.(2); PIEROTTI J.(2); STUTT K.(2)
(1) Altria Client Services LLC, Richmond, VA, USA; (2) Eurofins Lancaster Professional Scientific Staffing, Richmond, VA, USA

In March 2018, the US Food and Drug Administration (FDA) issued an advanced notice of proposed rulemaking (ANPRM) for a nicotine product standard that would limit the amount of nicotine in combusted tobacco products. FDA requested comments regarding setting maximum nicotine levels in tobacco filler at 0.3 to 0.5 mg/g. FDA is also considering if this proposed product standard should specify a method for manufacturers to use for the determination of nicotine in very low nicotine content (VLNC) tobacco fillers. The ANPRM states that CORESTA developed a standardized method for the analysis of nicotine in unburned tobacco and tobacco products (CRM 62). However, CRM 62 has not been demonstrated to be fit for purpose for the analysis of tobacco fillers with nicotine levels in the range of 0.3 to 0.5 mg/g. Similarly, the ANPRM mentions that ISO 10315 is a standardized method that was developed for the analysis of nicotine in mainstream smoke, but as is the case with CRM 62, ISO 10315 has not been demonstrated to be fit for purpose for the analysis of VLNC smoke condensates. Our efforts were focused on lowering the limit of quantitation (LOQ) for these two methods. We will present the method modifications and validation results for CRM 62 and ISO 10315 for the determination of nicotine in VLNC tobacco filler and smoke condensate. The limit of quantitation for the modified methods was determined to be 0.02 mg/g and 0.01 mg/cigarette for CRM 62 and ISO 10315, respectively.

Urinary leukotriene E4 and 2,3-dinor thromboxane B2 are potential biomarkers of effect for tobacco product evaluations

MAKENA P.; LIU Gang; NOURI N.; CHEN P.; PRASAD G.L.
RAI Services Company, Winston-Salem, NC, USA

Adverse health outcomes from cigarette smoking generally take several decades to develop. Physiologically-relevant biomarkers of effect (BioEff), such as FEV1, typically require sustained smoking abstinence over several months to detect a significant change. However, short-term BioEff can be valuable tools for evaluations in tobacco product-switching studies, as they provide insights about early biochemical changes associated with pathophysiology of the health effects of the tobacco products being evaluated. Data from cross-sectional studies revealed that several arachidonic acid (AA) metabolites are useful BioEff, as evidence indicates changes in AA metabolites after switching from combustible tobacco products to noncombustible tobacco products. Here, in a series of smoking cessation and product switching clinical studies, we evaluated several urinary AA metabolites as potential BioEff. Pre- and post- tobacco product switching samples from internal clinical studies were collected and analyzed to quantify the levels of AA metabolites including prostaglandin (t-PGDM, t-PGEM, and PGF2α), isoprostane (iPGF2α-III, 8-iso-PGF2α, and 2,3-d-8-i-PGF2α), thromboxane (2,3-d-TXB2 and 11-d-TXB2) and leukotriene (LTE4) synthesis. Switching cigarette smokers to either cessation, oral tobacco products (snus or dissolvables), or electronic nicotine delivery systems (ENDS) resulted in significantly reduced levels of 2,3-dinor-TXB2 and LTE4. Our results demonstrated that 2,3-dinor-TXB2 and LTE4 represent potential BioEff in short-term tobacco product switching studies. 2,3-dinor-TXB2 and LTE4 play a role in platelet aggregation and inflammation, respectively, which contribute to cardiovascular and lung disease. Thus, 2,3-dinor-TXB2 and LTE4 may prove to be useful short-term BioEff in tobacco product evaluations of biological response.

Tobacco products – a cytotoxicity and genetic toxicology testing paradigm

BRUCE S.; DAKOULAS E.; DUTTA A.; PANT K.; ROY S.; YOUNG M.
MilliporeSigma, Rockville, MD, USA

As guidance is pending on the required testing for tobacco products, BioReliance Toxicology Testing Services® of MilliporeSigma as a contract research organization has worked with various tobacco companies to test their products using different cytotoxicity and genetic toxicology endpoints. We have tested a variety of tobacco products in different formulations e.g., combustible tobacco products (total particulate matter (TPM) – 46 versions, cigarette smoke aqueous extract (CSEaq) – 5 versions), flavorings (4 versions), casing solutions (1 version), e-liquids (19 versions), and tobacco heating products (3 versions). By using the standard regulatory genotoxicity test battery which is used for testing pharmaceutical and chemical entities, we can assess the safety of various types of tobacco products. The bulk of the testing thus far, in our laboratories, has been performed using in vitro assays like Neutral Red Uptake (NRU) with CHO-WBL and BALB/c 3T3 cells (19 studies), bacterial reverse mutation (Ames) assay (19 studies), in vitro micronucleus (MN) assay in CHO-K1 cells (13 studies) or TK6 cells (6 studies) and Bhas42 cell transformation assay in both initiator and promoter versions (51 studies). Additionally, in vivo comet assays (11 studies) have been performed by nose-only inhalation using both male and female rats. A high throughput in vitro flow cytometry-based CANFlow assay (Clastogenicity, Aneugenicity, and Non-genotoxicant) in TK6 cells can be used as an initial non-GLP screening assay to differentiate between genotoxic and non-genotoxic tobacco products and to classify genotoxicants as clastogens or aneugens. In summary, various tobacco products can be effectively evaluated for the genotoxic mode of action by a variety of assays for safety assessment in efforts to meet the regulatory requirements.

Do levels of menthol in mainstream cigarette smoke influence dependence? Data from NHANES 2007-2014

PARMS T.(1); VAN LANDINGHAM C.(2); MARANO K.M.(1)
(1) RAI Services Company, Winston-Salem, NC, USA; (2) Ramboll, Monroe, LA, USA

A 2013 preliminary scientific evaluation by the U.S. Food and Drug Administration (FDA) concluded that menthol in cigarettes is likely associated with increased initiation and progression to regular cigarette smoking, increased dependence, and reduced success in smoking cessation, especially among African American menthol smokers. However, there has been considerable debate as to whether the available evidence, in fact, supports these preliminary conclusions. Further, published studies analyzing whether varying levels of menthol in commercial cigarettes affect these smoking behaviors is not available. This paper addresses this gap in the literature by evaluating two indicators and a validated measure of cigarette dependence (i.e., cigarettes per day, time to first cigarette, and heaviness of smoking index [HSI] scores) among current regular smokers of cigarettes containing varying levels of menthol. Data from the 2007-2014 National Health and Nutrition Examination Surveys (NHANES) were analyzed. Smokers were defined as adults, ages 20 years and older, who had smoked ≥100 cigarettes in their lifetime, currently smoke (every day or some days), and exclusively used cigarettes in the last five days. Mean menthol content in mainstream smoke for various cigarette brands and styles was obtained from Altria Client Services' publically available response to FDA's Docket No. FDA-2010-N-0001, 75 FR 28027, and product names were matched in NHANES. Results indicated that after adjusting for age, gender, and race, there were no statistically significant differences in dependence as measured by cigarettes per day, time to first cigarette, or HSI by level of menthol in cigarette smoke. These findings - that the amount of menthol in mainstream smoke does not influence these metrics of smoking dependence - contribute to the evidence base regarding the potential impact of menthol in cigarettes on public health, and should inform regulatory decision making.

Reduction in urinary and blood biomarkers of tobacco exposure in smokers switched to an electronic nicotine delivery system product

KANOBE M.; JONES B.A.; CHEN Xiaoming; ECKHARDT S.; DARNELL J.; BROWN B.G.
RAI Services Company, Winston-Salem, NC, USA

Electronic nicotine delivery system (ENDS) products have the potential to provide nicotine to tobacco consumers while reducing exposure to combustion-related toxicants. This study evaluated changes in biomarker(s) of exposure (BOE) to combustion-related toxicants after smokers switched from their usual brand (UB) non-menthol cigarettes to one of three Vuse Original ENDS products (Vuse Vibe, Vuse Ciro and Vuse Solo) or to abstinence. The toxicants evaluated are identified by the FDA as harmful or potentially harmful constituents in tobacco smoke and listed in the ENDS Premarket Tobacco Product Application Draft Guidance for determination in e-liquids and aerosols. Subjects smoked their UB cigarette ad libitum for two days, then were randomized to one of three Vuse ENDS products for a 5-day ad libitum use or to abstinence. Twelve urinary (including nicotine and five metabolites) and one blood BOE were determined at both baseline and Day 5. Changes in BOE from baseline to Day 5 were determined and the percent change of mean calculated for each cohort. Results showed an overall decrease in BOE across the three product-switch and abstinence cohorts. Blood carboxyhemoglobin decreased 58-70% in all four study cohorts. Urinary biomarkers decreased between 40-95% for aromatic amines, mercapturic acids, tobacco-specific nitrosamines, and 3-hydroxy-benzo[a]pyrene. However, smaller changes (5-30% decrease) in total nicotine equivalents were observed from baseline to Day 5 for each of three product-switch cohorts versus a >95% decrease in the abstinence cohort. Nicotine levels within the product-switch cohorts remained relatively stable throughout the study. In this study, smokers who were switched to the ENDS products had reductions in combustion-related BOE similar to the reductions observed in the abstinence cohort.

Challenges in the administration of visual analog scales (VAS) during clinical pharmacology trials of ENDS products

COPELAND D.; RUSCH L.; FEIN M.; PEARSON M.
High Point Clinical Trials Center, High Point, NC, USA

Visual Analog Scales (VAS) are an important and commonly used tool for subject feedback in clinical trials. VAS scales subjectively measure characteristics across a range of values which cannot be easily measured by objective techniques or mechanical tools. This subjectivity is important to clinical trial outcome but requires consistently applied processes to ensure the data obtained can be compared across subjects. This research project demonstrates types of problems encountered. This research project demonstrates the benefits of training, supervision of the application and utilization of VAS to improve reliability and accuracy. Research and interviews with experienced Principal Investigators provided insights as to the use, application and results from VAS measurements. The factors most often cited affecting reliability of data, are the variability and inconsistency in completion of the scales. The research identified instructions often poorly written and/or misunderstood or misinterpreted by subjects, as well as time pressures for completion. Examples of problems in completion include subjects marking the scales erroneously, multiple responses on the same scale, creative illustrations, extraneous marks outside the scale ranges and incompletion of the scales. These types of responses may prevent accurate measurements and result in lost data. The issues may be compounded by failure to perform real time supervision during scale administration. Finally, we present a recommended process to ensure the highest degree of accuracy and representative subject feedback. By implementing these concrete steps, sites can ensure more accurate and consistent subject reporting on VAS measurements. In conclusion, it is important that training, instruction clarity and real time staff monitoring of completion by subjects is consistently applied during clinical trial execution to ensure consistent and interpretable results.

Sensitivity analysis of exposure parameters in quantitative risk assessment for substantial equivalence evaluation

PARMS T.; LIU C.; MARANO K.M.
RAI Services Company, Winston-Salem, NC, USA

Quantitative risk assessment (QRA) can provide a useful and efficient approach to address questions that might arise regarding human health risk and potential influence of tobacco products on public health. In evaluation of substantial equivalence (SE) applications, differences in harmful and potentially harmful constituents (HPHCs) may necessitate the determination of whether a new product raises different questions of public health; thus results from QRA are a valuable metric. This study is to evaluate impacts of several important exposure parameters, such as consumption rate, body weight, and lifespan, on estimated long-term health risks of combustible cigarettes and smokeless tobacco products in the context of SE evaluations. Different values of consumption rates, body weight, and lifespan are obtained from the most recent Centers for Disease Control and Prevention reports, National Health Interview Survey, the 1999-2016 National Health and Nutrition Examination Survey, US Environmental Protection Agency (USEPA) and US Food and Drug Administration recommended values, and the literature. Potential health risks are evaluated following a standard QRA process, utilizing toxicity values from regulatory sources and standard default values for other exposure factors. Results of the study indicate that tobacco consumption rates exert the most influence on estimated long-term health risks from exposure to HPHCs in tobacco products. Recent updates on body weight and lifespan values recommended by the USEPA have minimal impacts on predicted long-term health risks of tobacco users compared to previously recommended values. Regardless of the exposure parameters values, given the comparative assessment of risk between the new and corresponding predicate products in the context of an SE evaluation, the same input values would be applied for both products. Thus, the resulting difference between the estimated risk for each product would not be affected.

Differential gene expression and functional responses in peripheral blood mononuclear cells pretreated with combustible and non-combustible tobacco products

TRAN Q.T.(1); ARIMILLI S.(2); LIU Gang(1); MAKENA P.(1); PRASAD G.L.(1)
(1) RAI Services Company, Winston-Salem, NC, USA; (2) Wake Forest School of Medicine, Winston-Salem, NC, USA

Chronic smoking is known to alter immune responses, which compromises tumor surveillance and host defense against microbial infections. We previously showed that smokers (SMK) exhibit distinct peripheral blood mononuclear cell (PBMC) gene expression patterns relative to moist snuff consumers (MSC) and non-tobacco consumers (NTC). Moreover, exposure of PBMCs to cigarette whole smoke-conditioned medium (WS-CM) has been shown to suppress the expression of genes involved in inflammatory signaling pathways. We investigated whether PBMCs pre-treated with WS-CM or smokeless tobacco extract (STE) could activate signaling pathways in response to lipopolysaccharide (LPS) and phorbol 12-myristate 13-acetate (PMA)/ionomycin stimulation. Genome-wide expression was measured in PBMCs pre-treated for 3 hours with non-cytotoxic doses of WS-CM (low, medium and high doses) or STE (single dose); WS-CM and STE dosing was based on equi-nicotine units. Pre-treated cells were stimulated with LPS and PMA/ionomycin for 5 hours. The pre-treatment with STE and the low dose of WS-CM did not significantly alter the stimulated gene expression. However, cells pre-treated with the medium or high dose of WS-CM demonstrated many significant changes in LPS- and PMA/ionomycin-stimulated gene expression compared to un-treated cells. These changes were mostly a downregulation, indicating immune response suppression. The suppressive effects of WS-CM on inflammatory pathway-relevant gene expression persisted for over 24 hours, as evidenced by decreased levels of IFN , TNFα and IL-2 in stimulated cells. Pathway analysis also revealed that even after pro-inflammatory stimulation, higher doses of WS-CM continued to suppress many signaling pathways associated with immune cell differentiation and inflammatory responses. Our results provide a mechanistic understanding of how tobacco smoking, rather than smokeless tobacco use, compromises immune function, rendering SMK to be more susceptible for certain diseases.

Determination of the levels of short chain organic acids in tobacco and liquids used in electronic cigarettes

MOLDOVEANU S.; KILBY K.B.; HUDSON A.G.
R.J. Reynolds Tobacco Company, Winston-Salem, NC, USA

Short chain organic acids from acetic to caprylic play an important role in tobacco flavor, and may be present in the flavors added to the liquids used in electronic smoking devices (e-liquids). For this reason, the determination of the level of those acids is of interest. Until recently, the GC/MS analysis of these acids has been problematic due to the reactivity of the analytes within the column phase that lead to the lack of reproducibility of results. The use of a new GC column, DB-624UI from Agilent leads to much better reproducibility. In this study, the column has been used for the separation and GC/MS analysis of several short chain organic acids including acetic, propionic, iso-butyric, butyric, iso-valeric, valeric, 3-methylpentanoic, caproic, heptanoic and caprilyc in tobacco and e-liquids. The analysis in e-liquids can be done directly following the dilution of the sample in methanol and injection in the GC/MS instrument. However, the direct analysis of the acids in tobacco after extraction, for example with methanol or acetonitrile, shows significant interferences. For this reason, a solid phase extraction (SPE) step is necessary to clean up the extract. This is done with retention from an acetonitrile extract on a Strata SAX 500 mg/3mL SPE cartridge, and elution with methanol containing 5.0 mg/mL formic acid. The eluted samples generated very reproducible results. A set of 18 tobaccos and six different e-liquids were successfully analyzed for short chain organic acids by this procedure.

Pyrolysis of glycerin at different temperatures and heating rates

MOLDOVEANU S.; CHOU Ching-En
R.J. Reynolds Tobacco Company, Winston-Salem, NC, USA

Glycerin is one of the main components of the liquids used in electronic cigarettes (e-liquids), and during their use, the e-liquid is exposed to increased temperatures. Information about glycerin pyrolysis is available in the literature (see e.g. R.R. Baker, L.J. Bishop, J. Anal. Appl. Pyrol., 71 (2004) 223-311, or S.C. Moldoveanu, Pyrolysis of Organic Molecules with Applications to Health and Environmental Issues, Elsevier, Amsterdam, 210, pp.284-286), but this information covers only a limited temperature range, typically higher than the temperatures attained in an e-cigarette. For this reason, a more detailed study on pyrolysis of glycerin at different heating conditions was considered necessary. Glycerin (containing about 10% water) was subjected to pyrolysis at five different temperatures in the range 350 °C to 750 °C, at two different heating rates 20 °C/ms and 5 °C/ms. Also, a mixture 50/50 w/w of glycerin and silica gel (Selecto Scientific Ga, USA) 100-200 mesh was subjected to pyrolysis at heating rate of 20 °C/ms at the same range of temperatures. Results for the dependence on temperature of the generation of compounds such as 1-propene, formaldehyde, acetaldehyde, 2-propenal, acetic anhydride, 2-propen-1-ol, lactic acid, acetol, glycidol are reported, showing that most of these compounds are formed only at higher temperatures, while they are absent in the range 350 °C to 450 °C.

One more time: unprotonated nicotine in e-cigarette aerosols: is it really there?

LAUTERBACH J.H.
Lauterbach & Associates, LLC, Macon, GA, USA

A May 23, 2018, article in Chemical Research in Toxicology claimed, "E-liquids designed to combine high total nicotine level (addictive delivery) with low αfb (for ease of inhalation) are likely to be particularly problematic for public health." In the quoted sentence, high total nicotine level would be in the range of 50 to 60 mg/mL. αfb is the fraction of total nicotine that is protonated. A low-level of αfb would be 0.05, and a high-level would be 0.80. The values refer to fraction unprotonated nicotine in undiluted e-liquids and were determined with proton NMR spectroscopy. However, the NMR instrument cannot taste, smell, or sense irritation. Moreover, humans do not inhale unaerosolized e-liquids or condensed aerosols (aerosol collected matter) from machine vaping of e-cigarettes. A more appropriate and less costly technique for estimating the pH of the aerosolized e-liquid is to use a glassmouth with artificial saliva in the bottom of the mouth and a pH-electrode in the aerosol stream as generated the puffing conditions specified in CRM 81, as has been previously reported. Aerosol pH-values for the following devices and e-liquids (all in 50 mg/mL nicotine in PG with V2 EX cartomizers): 2-hole rectangular, 5.71; 4-hole cylindrical, 7.40; 10 mg/mL menthol, 8.27; 10 mg/mL vanillin, 7.64; 20 mg/mL benzoic acid , 7.12; and 59 mg/mL quinic acid, 7.32. Corresponding saliva pH-values were 6.69, 7.11, 7.75, 7.60, 7.08, and 7.08. In addition, device configuration and water content of the aerosol are important in determining the potential for unprotonated nicotine to partition into the gas-vapor phase of the aerosol. When those factors are taken into account, the partitioning unprotonated nicotine in to the aerosol is still minimal.

Evaluation of the formaldehyde hemiacetals and acetals relevant to electronic cigarettes

JIN Xiaohong Cathy; AVERY K.C.; BALLENTINE R.M.; GARDNER W.P.; McKINNEY W.J.; MELVIN M.S.; PITHAWALLA Y.B.; SMITH D.C.; WAGNER K.A.
Altria Client Services LLC, Richmond, VA, USA

In 2016, the Food and Drug Administration (FDA) issued draft Guidance to Industry regarding the submission of Premarket Tobacco Applications for Electronic Nicotine Delivery Systems (ENDS). In this draft guidance, the FDA recommends reporting toxicants in aerosols, including formaldehyde. Recent publications suggested that 1) formaldehyde levels in ENDS products were underreported because formaldehyde may react with propylene glycol (PG) and glycerin (Gly) in the aerosol to form hemiacetals; 2) the equilibrium would shift from the hemiacetals to the acetals in the acidic DNPH trapping solution. In both cases, neither the hemiacetal nor the acetal would react to form the target formaldehyde hydrazone due to the lack of the carbonyl functional group; thus, underreporting formaldehyde. These reports were studied in our laboratory. Our results showed that the aerosol generated from formaldehyde spiked e-liquids gave near quantitative recovery of formaldehyde in the aerosol suggesting that if any hemiacetal was formed in the aerosol it would readily hydrolyze to free formaldehyde in the acidic DNPH trapping solution. We demonstrated that a custom synthesized Gly-hemiacetal added to DNPH trapping solution would readily hydrolyze to form the formaldehyde hydrazone. We also demonstrated that acetals of PG and Gly added to DNPH trapping solution would not hydrolyze to form the hydrazone. In order to further investigate the reported presence of acetals in e-liquids and aerosols, we developed a qualitative GC-MS method for the identification of the PG and Gly acetals of formaldehyde. No acetals of PG/Gly were detected in the e-liquids or the aerosols. These results demonstrate that formaldehyde levels in e-cigarette aerosols are not underreported due to hemiacetal and acetal formation.

Survey of carbonyls in aerosols from electronic nicotine delivery systems

CULBERT C.G.(1); WEIDMAN R.A.(2)
(1) RAI Services Company, Winston-Salem, NC, USA; (2) R.J. Reynolds Tobacco Company, Winston-Salem, NC, USA

Due to the growth of the electronic nicotine delivery systems (ENDS) category and the potential role of ENDS in switching and cessation from traditional combustible cigarettes, there is increased interest in the relative yields of harmful and potentially harmful constituents (HPHCs) that can be expected from ENDS and combustible cigarettes. In this study, closed ENDS products were analyzed for carbonyl yields in their aerosols. The carbonyls evaluated included formaldehyde, acetaldehyde, crotonaldehyde, acrolein, diacetyl, and acetyl propionyl from the recommended list of HPHCs in the FDA draft guidance on Premarket Tobacco Product Applications for ENDS. Cig a like, tank, and pod mod closed ENDS products were evaluated to survey a range of possible closed product configurations. To account for the range of flavors available for these types of products, tobacco, menthol, and fruit flavored cartridges were evaluated for each of the products. Aerosols were generated using the 55 mL puff for 3 seconds every 30 seconds with a square wave puff profile from CORESTA Recommended Method No. 81, as well as a more intense puff regime involving an 80 mL puff for 5 seconds every 15 seconds with a square wave puff profile. These results were compared to a survey of combustible cigarette carbonyl yields generated under both ISO and Health Canada Intense puff regimes. ENDS carbonyl yields were generally found to be lower than the corresponding combustible cigarette carbonyl yields.

Characterization of a temperature regulated e-cigarette for potential use as a reference device

PENNINGTON A.; GILLMAN I.G.
Enthalpy Analytical, Durham, NC, USA

The development and adoption of a well characterized reference e-cigarette would be useful during the routine analysis of e-cigarette emissions. One possible class of devices that might be suitable for use as a reference e-cigarette are temperature regulated devices. Temperature regulated devices control heating of the coil system by measuring changes in coil resistance during use. Temperature control may limit the production of thermal degradation products (TDPs) and dry puffing by the user. In this study, we used 20 commercial devices fitted with a nickel coil inside a refillable tank. Devices were analyzed for total aerosol yield, and the formations of aldehydes. Devices were tested over a coil temperature range of 201°C to 316°C. A temperature of 300°C was chosen for evaluation of device reproducibility and repeatability. Devices were found to have a reproducibility of 20-37% RSD in aerosol mass generation and 8-70% RSD for formaldehyde/gram. Devices produced 20-300, 5-100, and 2-10 ug/gram of formaldehyde, acetaldehyde, and acrolein respectively. Furthermore, only 50% of devices used were found to be reproducible with a RSD within 20%. Devices were found to have a repeatability of 33% RSD in aerosol mass generation and 84% RSD for formaldehyde/gram. Devices produced 30-600, 10-300, and 2-40 ug/gram of formaldehyde, acetaldehyde, and acrolein respectively. Transfer efficacy studies were performed at a lower temperature (231°C) with a 50:50 PG:VG 2% nicotine e-liquid, spiked with formaldehyde and acetaldehyde. Devices showed good reproducibility with RSDs of 8.9% and 9.1% on a per gram basis for formaldehyde and acetaldehyde transfer. Results from this study showed that temperature regulated devices cannot be utilized to precisely produce thermal degradation products, but can be used as an analytical reference to accurately assess the delivery of aldehydes from liquids that contain known amounts of these compounds.

Structure-activity relationships of propylene glycol, glycerin, and select analogs for carbonyl thermal degradation products

MELVIN M.S.; BALLENTINE R.M.; GARDNER W.P.; McKINNEY W.J.; PITHAWALLA Y.B.; SMITH D.C.; WAGNER K.A.
Altria Client Services LLC, Richmond, VA, USA

In 2016, the Food and Drug Administration (FDA) issued draft Guidance to Industry regarding the submission of Premarket Tobacco Applications for Electronic Nicotine Delivery Systems. In this draft guidance, the FDA recommends reporting several carbonyl compounds; including formaldehyde, acetaldehyde, and acrolein, whose presence in electronic cigarette aerosols is typically attributed to thermal degradation of propylene glycol (PG) and/or glycerin (GLY). Studies were conducted using carbon-13 labeled PG and GLY to elucidate the mechanisms and sources of the specific carbonyl compounds. These studies utilized a model reaction system based on microwave heating. GLY was found to be the primary source of formaldehyde while PG was found to be the primary source of acetaldehyde and acrolein. Formaldehyde and acetaldehyde generation from GLY is typically attributed to the retro-aldol condensation of 3-hydroxyproponal. This reaction was studied in further detail utilizing the microwave model system. These results were used to design PG and GLY analogs containing functional group substitutions to block reactive sites on the PG and GLY analogs. Blocking the reactive sites was anticipated to reduce the thermal decomposition of PG and GLY and thus provide information on the critical reaction centers required for the formation of the observed products. The results of these studies will be reported as well as the carbonyl yields from these PG and GLY analogs.

The relationship between water content and relative humidity for e-liquids

ST. CHARLES F.K.
St.Charles Consultancy, Lewisville, NC, USA

The water content of two e-liquids with glycerin to propylene glycol ratios (G/P) of 1.6 and 9.5 to 1 was determined at relative humidities ranging from 27% to 75%. The measured values were compared to theoretical values which were calculated from published data for glycerin and propylene glycol. The calculated values were close enough to those measured that the calculation should suffice for any future determinations of water versus relative humidity for e-liquids made with these humectants. The largest difference in measured and theoretical water content was at 75% relative humidity (G/P = 1.6), where the theoretical 38.8% water was 1.6% higher than the measured value. All other differences between measured and theoretical values were less than 1% water. For the G/P of 1.6, the root-mean-square (RMS) difference between measured and calculated values was 0.9% water and differences generally increased with increasing water. For the G/P of 9.5, the RMS difference was 0.6% water with a maximum difference of 0.8%. The calculation method is described. Included are equations describing water versus relative humidity for the two humectants.

Are all FSC banding agents inert?

LAUTERBACH J.H.
Lauterbach & Associates, LLC, Macon, GA, USA

Banding agents used on FSC (Fire Standards Compliant) cigarette paper are typically composed of polysaccharides, fillers, and additives. Polysaccharides used in commercial FSC papers include cellulose, sodium alginate, and modified starch. These banding agents reportedly act by modifying the diffusivity of the cigarette such that there are unfavorable conditions for propagation of the smoldering cigarette coal through the banded area. Most KS filter cigarettes have two bands and some 100-mm cigarettes have three bands along the cigarette rod. It also has been assumed that the mainstream smoke (MSS) stream from the lit end of the cigarette has no effect on the properties of the bands towards the filter-end of the cigarette. However, TPM, nicotine, water and carbon monoxide data from cigarettes made with FSC paper versus their non-FSC counterparts indicated that some types of FSC paper may be affected by the MSS resulting in decreased band diffusivity. Consequently, MVA Scientific Consultants (Duluth, GA) were retained to provide FT-IR, SEM, and EDX analyses of the banded and nonbanded regions of cigarette paper taken from commercial cigarettes. While cellulose and alginate bands were very uniform, bands formed from modified starch showed starch granules, partially disrupted starch granules, and soluble starch. When KS cigarettes made with starch-banded paper were partially smoked, and the rear band examined with optical microscopy after staining with Lugol's iodine solution, it was found that the the rear bands had a much smoother surface than the bands from the paper taken from unsmoked cigarettes. This finding supports the hypothesis that starch granules are being gelatinized from the water in the MSS, and the gelatinized starch is occluding the pores in the banded areas.

Functional filter wrapping materials and their impacts on specific properties of factory made cigarettes and heated tobacco products

LINDNER M.
TANNPAPIER GmbH, Traun, Austria

Filter wrapping materials are a synonym for mouthpiece covering sheets which include Tipping and filter plug wrap paper. Both sheets play a highly essential role for the manufacturing of conventional filter cigarettes and heat-not-burn items as they connect the filter plug with the tobacco rod, control the level of smoke constituents through permeability and shape the geometrical dimensions of smoking articles. In the present study, additional functionalities of filter wrapping materials will be demonstrated together with their potential application for factory made cigarettes (FMC) and heated tobacco products (HTP) and their respective influences on selected product features. Hereby, the individual functionalities are based on tactile, haptic, sensory and physical effects, in which the latter are focused on smoke yields and temperature reducing material properties. In order to enhance the haptic interaction between the consumer and the FMC / HTP as well as the tactile comfort during the smoking process, mechanical embossing of Tipping Paper, a velvety surface coating and a highly hydrophobic lip-release varnish will be discussed. Since Tipping Paper represents a non-burnt component of FMCs / HTPs, it will be shown that flavors can be applied for the stimulation of the primary tastes inside the oral cavity without toxicological concerns from pyrolytic side products. And finally, the application of active substances on filter plug wrap paper will be introduced with the purpose to reduce gaseous deliveries and the thermal energy of the HTP smoke. The efficiency of these active agents will be investigated under pilot plant test conditions and underlined by quantitative results.

A study on the transfer of cigarette paper ingredients into mainstream smoke

WANNA J.(1); CHEN Cai(2); WANG Xiaoping(3); XIAO Cuicui(3)
(1) SWM, Alpharetta, GA, USA; (2) SWM Shanghai Representative Office, Shanghai, P.R. China; (3) China Tobacco Mauduit, Jiangmen, Guangdong, P.R. China

To study the influence of cigarette paper on mainstream smoke, three ingredients, levulinic acid (LA), ethylvanillin glucoside (EVG) and encapsulated ethylvanillin (EnEV) were added into cigarette paper as surrogate tracers, and king-size (KS) and superslim (SS) cigarettes were produced. Smoke deliveries of these cigarettes were measured under the ISO smoking regime. Amount of nicotine and the tracer compounds LA and ethylvanillin (EV) were calculated both in mainstream smoke and in filters. The results indicate that: 1) the transfer rates of the tracer compounds from paper to smoke are in the order of LA > EnEV > EVG; 2) ratio of the total amount of tracer compounds to nicotine (LAsmoke+filter/nicotinesmoke+filter, EVsmoke+filter/nicotinesmoke+filter) from the SS cigarettes are significantly higher than those from KS cigarettes; 3) ratio of the tracer compounds between smoke and filter (LAsmoke/LAfilter and EVsmoke/EVfilter) are much lower than the ratio of nicotine between smoke and filter (Nicotinesmoke/Nicotinefilter). These results provide more evidence supporting the theory that, comparing to tobacco filler, ingredients in cigarette paper may contribute differently into smoke due to their chemical/physical properties and the position in the outer surface of a cigarette. They also suggest that cigarette paper has more significant influence on cigarettes with reduced circumference than KS cigarettes.

The impact of waterpipe configuration and smoking regime on the concentration and size distribution of waterpipe tobacco smoke

HAUSER C.D.; CARMACK D.; MAILIG R.; ONIA T.; BERND K.
Davidson College, Davidson, NC, USA

The combination of shisha as a very different matrix from cigarette tobacco and indirect heating versus combustion may present unknown health hazards in waterpipe tobacco smoke (WTS). We are focused on how waterpipe use and design impact the physical properties of particles in WTS and thus their potential toxicity. A waterpipe can be viewed as a mini atmospheric system where vapors condense to form particles and particles increase or decrease in size or number due to coagulation, agglomeration, condensation or evaporation. Not only do the constituents of shisha, such as glycerol, syrup and tobacco, and heating method contribute to the formation of particles but waterpipe design elements such as pipe height and hose length could potentially impact aging processes. Here, we present the results of systematic investigations into how pipe height, hose length and user topography or smoking regime contribute to the distribution and concentration of the particulate component of WTS. Commercial waterpipes were smoked using the Beruit method in comparison to other methods with a TSI Engine Exhaust Particle Sizer (EEPS) sampling particles from 5.3 to 530 nm at rate of 10Hz. Particle distributions and concentrations are evaluated as a function of time. Preliminary investigations of the impact of pipe height and hose length indicate decreasing concentrations and decreasing mean particle diameter with increasing height and length. Not only do these results inform toxicity studies, but they also provide information critical to understanding the formation and aging of particles in WTS.

Are available test methods for the determination of ammonia in mainstream cigarette smoke fit for the analysis of cigars? – Part 2

PREPELITSKAYA Y.; MILLER J.H.; RUSYNIAK M.; HIGGINS F.S.; AVERY K.C.; CROSSWHITE M.; SPICER-GLAVE S.; SHAH N.
Altria Client Services LLC, Richmond, VA, USA

In May 2016, the U.S. Food and Drug Administration (FDA) issued a final rule deeming cigar products within scope of their regulatory authority. Under this regulation the FDA will require manufacturers to report the quantities of Harmful and Potentially Harmful Constituents (HPHCs) in cigar filler and smoke. Currently, FDA has not indicated that the list of constituents for cigars would differ from that of the list of constituents provided for tobacco smoke, and thus would require the analysis of ammonia in cigar smoke. We previously showed that the CORESTA Recommended Method No. 83 for the analysis of ammonia in cigarette smoke may not be fit for purpose for all cigar types. We previously presented, at TSRC in 2017, data that shows amount of ammonia measured in cigar smoke depends on tobacco blend composition, and on the concentration of acid in the trapping solution. We also showed that ammonia is not stable in the final extracts, specifically, when stored at room temperature. Our current work is focused on improving our understanding of the formation of ammonia in cigar smoke to better understand the differences between cigars made with different tobacco blends. We have used Fourier-transform infrared spectroscopy (FTIR) and high resolution mass spectrometry to investigate the amount and potential source of ammonia measured in smoke from different cigars blends. In addition, we evaluated the trapping efficiency using ammonia gas to identify the optimal conditions for trapping. These results show that the ammonia gas is sufficiently trapped on the Cambridge Filter Pad and in the first impinger containing 0.1N H2SO4. We will also present data for ammonia for a range of market products with high burley content and with high dark air-cured tobacco content.

Characterization of microbial communities associated with three smokeless tobacco reference products and correlation with chemical characteristics

MOE L.; GREENHUT I.; JI Huihua
University of Kentucky, Lexington, KY, USA

Our first objective is to characterize microbial community composition associated with loose leaf chewing tobacco, American Snus and Swedish Snus smokeless tobacco reference products (STRP) as part of the Center for Tobacco Reference Products at the University of Kentucky, and to identify changes in these communities that occur as a function of time and storage conditions. The secondary objective is to correlate changes in the microbiome with changes in the content of biologically active chemicals such as nicotine and carcinogenic nitrosamines. We used culture-dependent methods, specifically dilution plating to Tryptic Soy Agar, de Man Rogosa Sharpe medium, and Acidified Potato Dextrose Agar to quantify the culturable bacterial loads and yeasts present on all three products over the course of one year in storage at -20°C, 4°C, and 27°C, and over the course of 5 weeks at -20°C, 22°C, and 37°C. We also constructed a library of isolate strains for identification and further study. Subsequently, 16S rRNA gene amplicon sequencing was employed to characterize the bacterial microbiota associated with each STRP product at each timepoint in the storage studies. Statistical analysis revealed bacterial communities specific to each product. Changes in these microbiomes were further correlated with changes in physicochemical parameters. Further work is directed at correlating the enrichment of specific taxa and metabolic pathways with observed changes in the concentration of alkaloid-derived compounds in the STRPs over time and under different storage temperatures.

Comparison of sequencing platforms for microbial profiling in smokeless tobacco

LI Xingpeng; GERARDI A.; BROWN J.
R.J. Reynolds Tobacco Company, Winston-Salem, NC, USA

Advances in next-generation sequencing technologies have facilitated the analysis of microbial community composition through 16S rRNA amplicon sequencing. This results in greater throughput and enables identification of difficult-to-cultivate or uncultivatable bacteria. By using the Illumina platform, we sequenced different regions of 16S and obtained bacteria composition profiles in reference smokeless tobacco products CRP1.1, CRP2.1, and mock microbiome samples. Results from different 16S regions were compared to evaluate accuracy and consistency in bacteria composition profiling. In addition, we found that short sequencing reads often limit the bacterial composition characterization. This is due to the high similarity of 16S rRNA amplicon sequences across multiple genera when only comparing a selective region. To overcome this limitation we also used the Nanopore sequencing and PacBio Circular Consensus Sequencing (CCS) platforms to sequence full-length 16S rRNA amplicon libraries on the same reference smokeless tobacco products as well as mock samples. Both full-length sequencing platforms provide better taxonomic resolution than the Illumina short-read sequencing platform. A comparison of the Illumina, Nanopore, and PacBio 16S rRNA amplicon CCS results indicate that PacBio CCS provides the most accurate result on the mock samples and most consistent results on all samples tested. Therefore, PacBio CSS is the recommended choice for accurate and efficient characterization of microbial populations in smokeless tobaccos among 3 platforms tested.

Stability of certified smokeless tobacco reference products

GUAN Siqi; JI Huihua; BUSH L.P.
University of Kentucky, Lexington, KY, USA

In 2016, University of Kentucky, Center for Tobacco Reference Products (CTRP) was awarded a Cooperative Agreement with the US Food and Drug Administration (FDA) to produce certified smokeless tobacco reference products, including loose leaf (3S1), snus (1S5), and "Swedish" style snus (1S4). These smokeless tobacco reference products were manufactured in June 2017. The objective of this study is to evaluate the stability of smokeless tobacco reference products during long-term storage. 3S1, 1S5, and 1S4 were stored at room temperature (~22 °C), 4 °C, and -20 °C for 1, 2, 3, 6, 9 and 12 months. Some important parameters of smokeless tobacco reference products were determined at each sampling time, including moisture, pH, tobacco-specific nitrosamines (TSNAs), nicotine, free nicotine and benzo[a]pyrene (B[a]P). Prior to analysis, the reference products stored at -20 °C were transferred to a refrigerator for 24 hours and then moved to room temperature until they reached temperature equilibrium. The study indicated that the three smokeless tobacco reference products were stable in most parameters we measured, except that pH and free nicotine content continued to decrease at room temperature with the increased storage time. The amount of pH and free nicotine showed a moderate decrease at 4 °C while they were stable at -20 °C.

CTP submissions – tips and insight for preparing an electronic submission

SHOLTES D.(1); SMITH J.K.(1); ANGERMEIER G.(2)
(1) U.S. Food & Drug Administration, Silver Spring, MD, USA; (2) WiseDesign, McLean, VA, USA

Currently, tobacco product submissions are organized differently across application types and often, across applicants, and are received in a format that is not amenable for reviewers to organize and easily find what they need. This "unstructured" format is not conducive to automation and complicates analysis across applications and products. Familiarity with FDA's supported technical file formats, data specifications and standards will aid in the creation of an electronic submission of sufficient functionality to facilitate its processing and review by CTP. CTP intends to be consistent, wherever possible, with existing paradigms, file formats, and data standards developed by other FDA Centers. CTP strives to leverage existing electronic standards and participate in existing standards bodies. Both FDA and industry have benefited from the use of common technical standards including searchability, reliability, accuracy, and harmonization of submission structure across application types. Such standards also facilitate the development of supporting tools by the commercial market for regulatory submission creation, review, and analysis. FDA regulations avoid technical details on submission format and organization as technologies change over time. These details are outlined in technical specifications and guidance documents and adopted industry standards. Historically, independent standards associations have helped drive eSubmission standards and FDA has participated as a member. FDA can endorse such standards through Federal Register notice. FDA is currently evolving towards using the Health Level Seven (HL7) Regulated Product Submission (RPS) standard for its submissions. For CTP the structure is named the electronic Tobacco Technical Document or eTTD. CTP is testing a draft eTTD specification with software companies that provide eCTD solutions to regulated industry to ensure technical viability. Once validated, the eTTD will proceed for public comment.

Machine made cigar leaf and product workshop

STEVENS R.(1); BLACK R.(1); GILLMAN I.G.(2); SMITH J.(3); HARTLEY M.(4); LINDEGAARD T.(5)
(1) ITG Brands LLC, Greensboro, NC, USA; (2) Enthalpy Labs, Durham, NC, USA; (3) Altria Client Services LLC, Richmond, VA, USA; (4) Universal Dark Air-Cured Group, Oxford, NC, USA

The regulation of tobacco products continues to be a growing focus around the world. For example, in 2016, the FDA finalized a rule in the USA that extended its regulatory authority to all tobacco products, including e-cigarettes, cigars, hookah, and pipe tobacco. As different types of tobacco products are included in the scopes of regulatory frameworks, it becomes increasingly important to understand the similarities and differences of these products compared to other regulated products as they relate to agricultural practices, regulatory framework, and the state of the science and testing for these products. The Cigar Workshop will consist of two presentations. The presenters will come from diverse backgrounds and will focus on these topics: • Agricultural practices - Growing and harvesting of cigar tobaccos all over the world, including challenges, opportunities, and comparisons to conventional cigarette tobaccos • Regulatory and Scientific Challenges – Challenges for cigars compared to other product categories will be discussed with a focus from seed to finished product. Approaches to address Regulatory and Scientific challenges for cigars will also be presented. Following the presentations, there will be a brief panel discussion/Q&A, with audience participation, designed to explore and learn more about Machine Made Cigars and the opportunities to advance the science and understanding of these unique tobacco products and the impact on the current, and future, tobacco regulations. Secondly, there will be a series of stations set-up to provide "Hands-On" visual comparisons of cigar tobaccos, machine made cigar product comparisons, and egulatory & Science experts in Cigar products available for Q&A.

Comparison of hydrophilic interaction liquid chromatography and reversed-phase separation in untargeted LC/MS analysis of 3D cell cultures exposed to cigarette smoke and aerosols from a novel tobacco vapor product

TAKANAMI Y.; KITAMURA N.; ITO S.
Japan Tobacco Inc., Yokohama, Japan

Aerosols from a novel tobacco vapor product (NTV) generate lower levels of potentially harmful constituents than cigarettes. In vitro cell cultures are expected to exhibit early-phase biological responses to cigarette smoke; a comprehensive analysis of cellular extracts is required to understand these biological effects. We exposed commercially available human MucilAir™ organotypic bronchial epithelial cultures to cell-culture media containing cigarette smoke- or NTV vapor-bubbled solution or nicotine for 4 h. The cellular extracts were subjected to hydrophilic interaction liquid chromatography (HILIC) and reversed-phase liquid chromatography (LC) separation, and were then analyzed by high-resolution mass spectrometry. The obtained peaks from smoke- or nicotine-exposed samples were categorized using self-organizing maps. Seventy-three peaks out of approximately 3,000 peaks obtained by HILIC(+) separation were categorized into a group that displayed concentration-dependent on smoke but did not respond to nicotine exposure. No such groups were found in peaks by reversed-phase LC separation. The 73 peaks were subjected to partial least squares regression analysis; a model was constructed using the peaks from smoke-exposed cultures as explanatory variables and smoke concentration (as total particulate matter equivalent) as response variables. The predicted exposure concentration was calculated using the peaks from NTV-exposed cultures and was lower than expected, indicating that the cell cultures were exposed to low concentrations of cigarette smoke constituents other than nicotine. Because cigarette smoke constituents and their metabolites are extremely complex and difficult to identify using authentic compounds, untargeted metabolome analysis and statistical methods can yield data on the relationship between exposure and biological effects.

FTIR chemometrics applied to e-liquids quantitative analysis

HIGGINS F.S.; THORN E.K.; RUSYNIAK M.J.
Altria Client Services LLC, Richmond, VA, USA

As the electronic cigarette industry evolves, the need for new rapid analytical techniques also grows. New methods are need, both to monitor quality in a high throughput manufacturing environment and in a fast paced product development space. We report on the development and validation of a quantitative Fourier Transform Infrared spectroscopy (FTIR) method for rapid (two minute), simultaneous analysis of all the major components of e-liquids. E-liquids are commonly analyzed by Karl Fischer analysis for water, and gas chromatography for components such as nicotine, propylene glycol, vegetable glycerin, and menthol. Compared to the traditional techniques, the FTIR method is faster, less expensive, requires no sample preparation and eliminates the need for consumables, hazardous solvents, laboratory gases, and reduces instrument maintenance. Data analysis using chemometrics is automated and simple to implement. The pre-processing algorithms, such as mean centering, derivatization, and cross validation are powerful tools in chemometrics, which allow accurate calibration of analytes with multiple infrared absorbance bands. Our method is well suited for high throughput routine analysis, including quality control, shelf life studies, and research involving water uptake in e-liquids. We will present the results of a full method validation and compare the results to validated Karl Fisher and Gas Chromatography methods.

Adaptations of mainstream cigarette smoke methods to other product categories

JOZA P.; RICKERT W.; BAO Mingliang; SHARIFI M.
Labstat International ULC, Kitchener, ON, Canada

The requirements for mainstream emissions testing of products other than conventional cigarettes, continues to increase. Methods initially developed for the determination of mainstream cigarette emissions provide a good foundation for the analysis of next generation nicotine products including heat-not-burn (HnB) and e-vapour products, as well as other types of burning products (e.g. cannabis). Adaptations to extend the scope of these methods to other mainstream emissions, require demonstrating the performance characteristics of the method are unaffected. To demonstrate a change in the matrix did not impact extraction efficiency or the instrumental analysis of the analytes, either an artificial matrix or test sample matrix was used. The determination of lab fortified matrix recoveries, as well as the determination of resolution and tailing factor for chromatographic methods, were compared against conventional products. Trapping efficiency experiments were also performed to demonstrate the collection system was adequate for the collection of particulate phase and vapor phase analytes in these emissions. For the assessment of the analytical performance, multiple approaches were required depending on the product type, levels of the analyte of interest, and potential contributions from sources other than the test samples. Combinations of assessing laboratory reagent blanks, lab fortified blanks, lab fortified matrices and air blanks were required, depending on the specific scenario. Excluding any modifications or adaptations to enhance limits of quantification, methods developed for the analysis of mainstream emissions from cigarettes, were directly applicable to next generation nicotine products, as well as other types of burning products.

Comparison of potential health risks of combustible, heat-not-burn, and electronic cigarettes

LIU C.; MARANO K.M.
RAI Services Company, Winston-Salem, NC, USA

Given US Food and Drug Administration (FDA) regulation of tobacco products via the Family Smoking Prevention and Tobacco Control Act, there is a growing recognition of the need for evidence to assist both applicants and the FDA in making determinations of whether certain products may raise different questions of public health or are appropriate for the protection of public health. Quantitative risk assessment (QRA) is a useful tool to quantify relative public health impact. QRA was used to estimate cancer risk and cardiovascular, respiratory and reproductive or developmental toxic effects for conventional combustible cigarettes, heat-not-burn (HNB) cigarettes, and electronic cigarettes (e-cigarettes). Machine-generated yields of several harmful and potentially harmful constituents (HPHCs) specified by FDA in mainstream smoke or aerosol were obtained from the literature for representative products from the three different categories of tobacco products. Potential health risks were evaluated following the standard QRA process, utilizing toxicity values from regulatory sources and standard default exposure factors from the US Environmental Protection Agency and FDA, supplemented with assumptions to develop estimates of lifetime exposures. Emissions of HPHCs from HNB and e-cigarettes were considerably lower than those from cigarette smoke. Compared with conventional cigarettes and assuming comparable usage patterns, reductions in cancer risk were estimated to be 69% for HNB and 99% for e-cigarettes. In addition, reductions of 20% to 99% of cardiovascular, respiratory, and reproductive or developmental effects were estimated for HNB and e-cigarettes. These results demonstrate that estimated lifetime health risks are lower for HNB and much lower for e-cigarettes than those associated with smoking conventional cigarettes. Based on these findings, substituting conventional cigarette smoking with the use of HNB products or e-cigarettes will substantially reduce exposure to tobacco-specific toxicants and may reduce risk of cancer and noncancer health effects for the individual user.

Temporal variability of analytical testing for e-vapor products and impact on number of replicates

MORTON M.J.; GARDNER W.P.; AGNEW-HEARD K.A.; MILLER J.H. IV
Altria Client Services LLC, Richmond, VA, USA

In May 2016, the US Food and Drug Administration published a guide for industry "Premarket Tobacco Product Applications for Electronic Nicotine Delivery Systems." Among other things the guidance document provided testing recommendations to support Premarket Tobacco Application submissions for ENDS products, including a recommendation to generate data sets from a minimum of 10 replicates per batch and from a minimum of three batches for product analysis. To determine an appropriate number of replicates, it is important to consider the effect of the number of replicates on the uncertainty associated with the analytical result. For example, when all testing is conducted within a single laboratory, a component of the uncertainty of the analytical result is attributed to the temporal variability of the analytical method within that laboratory. That is, if two different samples are tested at two different times, temporal variability of the analytical method limits the precision of the comparison between the two samples. A consequence of temporal variability is that repeated testing in the short term has a rapidly diminishing effect on the uncertainty of the associated analytical measurement. Furthermore, if the standard errors of the analytical results are calculated without including temporal variability, the resulting standard errors can dramatically underestimate the actual uncertainty of the corresponding analytical results: Estimating the temporal variability of some analytical methods indicates that conducting 10 replicates in some analytes may be excessive.

E-vapor product appeal among tobacco users and non-users and the role of flavors in tobacco harm reduction

ZDINAK J.P.; VANSICKEL A.; GOGOVA M.; PLUNKETT S.
Altria Client Services LLC, Richmond, VA, USA

One of the key elements for tobacco harm reduction to be successful is to provide a portfolio of lower risk products that are appealing and can satisfy different segments or preferences of adult smokers. It is also important that the flavors do not appeal to non-users. We present results of two studies showing the need for e-vapor flavor varieties (EFVs) to appeal to ATUs and facilitate switching behavior. Study 1: The online Perceptions and Behavioral Intentions study enrolled ATUs (n=1,688) and non-users (n=1,252) to assess the impact of promotional materials on intentions and appeal of 14 EFVs. Across two conditions there was an approximately equal distribution of ATUs who selected the 14 EFVs as appealing, while the majority of non-users did not find the products appealing. Among non-users who did find a product appealing, results indicated that these individuals had high intentions to use at pre-test. Study 2: The 8-week, 2-phase, Actual Use study enrolled cigarette smokers not planning to quit (n=614) and e-vapor users (n=107). Phase 1: participants tried 14 EFVs (1/day) and rated liking of each. Phase 2: participants had open-access to the 14 EFVs for six-weeks and completed daily tobacco use surveys. All 14 EFVs were necessary for complete coverage of top-liking at trial. Adult smokers who completely switched had a larger choice set of EFVs than non-switchers and significant associations between number of EFVs used and decreases in cigarettes emerged. These findings support growing body of evidence indicating that EFVs can support the harm reduction potential of e-vapor products. FDA needs to carefully consider these findings in their assessment of the role of flavors on tobacco harm reduction.

Evaluation of benzo[a]pyrene (BaP) transfer from moist snuff to artificial saliva

MOLDOVEANU S.; MARSHALL J.; ST. CHARLES F.K.; POOLE T.
R.J. Reynolds Tobacco Company, Winston-Salem, NC, USA

Benzo[a]pyrene (BaP) has been listed by USFDA as one of 9 HPHC for smokeless tobacco. The compound may be present in moist snuff at levels between 50 and 200 ng/g (dry basis). However, in order to be transferred to a human user, BaP must be extracted first into saliva, and the compound has very limited solubility in water. Present study evaluated the "in vitro" extraction of BaP from moist snuff into artificial saliva. Three sets of experiments were performed on nine types of moist snuff: (1) extraction with water and measurement of BaP content of the water, (2) extraction with 6 portions of 20 mL water (120 mL) and measurement of the remaining BaP in the extracted snuff, (3) extraction with 4 portions of 20 mL artificial saliva followed by two portions of 20 mL water and measurement of BaP in the extracted snuff. The level of BaP in the water extract was below the detection limit. The level of BaP remaining in the water extracted moist snuff (after correction for the loss of components soluble in water) was 96.3% to 109.6% from the initial level of BaP, and the level of BaP in the extracted moist snuff with artificial saliva was 99.4% to 108.3% from the initial level of BaP. This "in vitro" study showed that in the conditions previously described, BaP remains entirely in the moist snuff matrix and it is not transferred into the artificial saliva (or water).

Method comparisons for particle size distribution and nicotine dissolution profiles in smokeless tobacco products

CUNNINGHAM C.K.(1); HANNA S.N.(1); KELLER C.L.(1); MEADOWS T.R.(2)
(1) RAI Services Company, Winston-Salem, NC, USA; (2) R.J. Reynolds Tobacco Company, Winston-Salem, NC, USA

In the ever-evolving regulatory environment in which tobacco manufacturers now operate, the need for innovation in regulatory research and method development is continual. One example of such research needs is a method for determining and comparing the tobacco particle size of products for regulatory purposes. If there are significant differences in particle size identified within the regulatory review process, scientists must justify these differences do not raise different questions of public health and as such, additional methods must be used and/or developed to respond accordingly. This work examined the particle size distributions of various smokeless tobacco products, including moist snuff tobacco and portioned moist snuff tobacco products. Additionally, nicotine dissolution was used to determine the correlation between particle size and rate of release of nicotine (if any) in a buffer solution. Particle size was determined via a sieve method utilizing multiple screen sizes in tandem to determine the particle size distribution of the tobacco products evaluated. Nicotine dissolution analysis was performed utilizing a dissolution apparatus 1 (App1; rotating basket) system under minimal rotation with an ammonium phosphate buffer solution (pH 7.4). Discussion will include statistical comparison of particle size distributions and corresponding dissolution profile comparisons, as well as dissolution analysis of portioned vs. non-portioned smokeless tobacco products. The relevance of particle size and surface area of smokeless tobacco products as compared to pharmaceuticals will also be discussed.

Toxicity value selection for quantitative risk assessment of harmful and potentially harmful constituents in smokeless tobacco products

AMANTANA A.; SAWANT S.G.; XIE Wei; LIU Shuang; MARANO K.M.; SZABO D.T.
RAI Services Company, Winston-Salem, NC, USA

The United States Food and Drug Administration (FDA) established a list of 93 harmful and potentially harmful constituents (HPHCs) in tobacco products and tobacco smoke, as required by the Family Smoking Prevention and Tobacco Control Act. These constituents are classified by FDA as carcinogens, respiratory, cardiovascular and/or reproductive/developmental toxicants. The abbreviated list of HPHCs for reporting in smokeless tobacco products includes acetaldehyde, arsenic, benzo[a]pyrene, cadmium, crotonaldehyde, formaldehyde, 4-(methylnitrosamino)-1-(3-pyridyl)-1-butanone (NNK) and n-nitrosonornicotine (NNN). The United States Environmental Protection Agency (EPA) published a hierarchical risk assessment approach to identify toxicity values for chemicals at Superfund sites. However, several toxicity values have not been updated for decades and may no longer be scientifically defensible. In an effort to conduct a human health quantitative risk assessment (QRA) on these HPHCs that could be used as inputs to assess risk associated with smokeless tobacco product consumption, a search of publicly available United States federal and state, and international databases for noncancer and cancer oral toxicity values was conducted. The search results included multiple values reported by different agencies. In the absence of FDA guidance for selecting a toxicity value for QRA, we report a selection approach supported by the most current scientific methods that include study quality, date of study, and toxicity value derivation approach. For noncancer toxicity values derived from a critical effect other than the endpoint(s) published by FDA, scientific justification was included to support the public health protective nature of the toxicity value(s) selected. This approach can be used to justify the selection of toxicity values for HPHCs identified by FDA in smokeless tobacco and other products for which the oral route of exposure is relevant.

An improved method for the determination of sodium propionate in tobacco products

WILLIAMSON F.; BAILEY J.
Global Laboratory Services, Wilson, NC, USA

Sodium propionate is a preservative with fungicidal properties that is often added to tobacco, primarily smokeless tobacco products, to prolong shelf life. Several regulatory agencies, including Health Canada and ANVISA, require the testing of sodium propionate in tobacco. While there is currently a method for the determination of sodium propionate in whole tobacco for Health Canada (T-312, December 1999) this uses GC-FID and an acidified organic solution for extraction that proves detrimental to the GC instrument resulting in front-end maintenance and cleaning after each run. This presentation describes the development and validation of a method for the determination of sodium propionate in tobacco and the challenges encountered. Sodium propionate was extracted from tobacco with hydrochloric acid, where it was converted to propionic acid, and partitioned into tert-butyl methyl ether (MTBE) with excess sodium chloride. An aliquot of the upper MTBE layer was subsequently analyzed by GC-MS. Separation of propionic acid from other components was achieved using a DB-WAX UI (30 m x 0.25 mm; 0.25 µm) analytical column and a temperature gradient (60 °C to 240 °C). Quantitation was performed using propionic acid-13C3 as an internal standard. Validation was performed on a range of different tobacco products including the reference products 2S1, CRP4, and CRP4.1, which list sodium propionate as an ingredient. Levels of sodium propionate (as propionic acid) observed in tobacco ranged from not detected to approximately 2900 µg/g for both 2S1 and CRP4. No sodium propionate was observed in CRP4.1.

Rapid measurement of phytochemicals in tobacco leaves using fluorescence fingerprint analysis

NAITO H.(1); TSUTA M.(2); KOIKE T.(1)
(1) Japan Tobacco Inc., Yokohama, Japan; (2) National Agriculture and Food Research Organization / Food Research Institute, Yokohama, Japan

Recently, progress has been made in rapid analysis of tobacco leaves. The amounts of water, sugar, and nicotine in tobacco leaves have been measured using near infrared spectroscopy. However, components that have low contents, such as phytochemicals, cannot be measured by near infrared spectroscopy because of its low sensitivity. Fluorescence fingerprint analysis has been applied to rapid analysis. Fluorescence fingerprint is a set of fluorescence spectra acquired at consecutive excitation wavelengths, creating a three-dimensional diagram. The pattern of this diagram is unique for every constituent, rather like a fingerprint. It is nondestructive and noncontact, inexpensive, and provides more information than near infrared spectroscopy. In addition, signals corresponding to intramolecular conjugated systems can be acquired with strong signal intensities and high sensitivity. It is expected that this method could be successfully applied to phytochemicals, which contain conjugated systems and are present at low levels in tobacco. In this study, fluorescence fingerprint spectroscopy was applied for quick and convenient analysis of phytochemicals in tobacco leaves. Seventy-two tobacco leaf samples were tested. First, the phytochemical contents were quantitated by chemical analysis. Then, the samples were pulverized, and front-face fluorescence spectra were measured using a F7000 fluorescence spectrophotometer (Hitachi High Technologies). Multivaliable analysis was used, and the correlation coefficient was 0.91 between the value measured by chemical analysis and predicted value obtained from the fluorescence fingerprint spectrum, and the root mean square error of prediction was 0.18 %. Our experiments will be discussed in detail on the day.

Characterization and certification of smokeless tobacco reference products

SLONE S.; CANETE S.; JI Huihua; JACK A.M.; BUSH L.P.; CHAMBERS O.; SHELTON B.
University of Kentucky, Lexington, KY, USA

The constituents of raw tobacco products are naturally variable, and product design, the manufacturing process and the evolving science of tobacco analytics may increase this variability. The call for certified tobacco reference materials and their potential application in product research and regulation indicates a need to understand this variability. The United States Food and Drug Administration awarded a Cooperative Agreement to the Center for Tobacco Reference Products at the University of Kentucky to produce and characterize four smokeless tobacco reference products including two snus, a moist snuff and a loose-leaf chewing tobacco. The statistical method for certification highlights the impact of measurement variability and laboratory differences on the certification process and therefore ultimately on the Certificate of Analysis. Characterization data from four ISO-accredited laboratories were used for the certification. Of the 14 parameters examined, acetaldehyde, formaldehyde, free nicotine, total nicotine, arsenic and benzo[a]pyrene exhibited larger than expected measurement variability, and crotonaldehyde was below the limit of quantitation for all samples. Graphics of the data and Cochran and Grubbs outlier analyses illustrate the types of variability experienced for the six constituents with large measurement variability, including consistent laboratory means with large between-lab variance and large repeatability standard deviations within a specific lab.

Development and verification of a black shank race 0 qPCR assay derived from RAPD marker assay

LESNIAK M.F.(1); SHEW D.(2)
(1) R.J. Reynolds Tobacco Company, Winston-Salem, NC, USA; (2) North Carolina State University, Raleigh, NC, USA

Historically, Ph gene coupling and repulsion RAPD marker assays (Johnson et al, 2002) have been used to screen for resistant and susceptible tobacco cultivars. Using them as a foundation for our work, DNA fragments from RAPD assays were extracted from corresponding bands in an agarose gel and sequenced. Sequences from each coupling and repulsion marker fragment were compared to homologue sequences in the available tobacco genome. Shorter sequences unique to resistant and susceptible cultivars were used for design of the probe based marker binding regions. Twenty-six separate parental varieties were screened with both the coupling and repulsion qPCR based markers, and yielded identical results to the RAPD assay. To further test the veracity of the newly-designed assays, a greenhouse screen was devised. It consisted of 12 resistant and 12 susceptible parentals, 12 F1's, and 250 F2's, grown under greenhouse conditions. Oat grains colonized by Phytophthora nicotianae were used to inoculate five to six-week old seedlings. One colonized oat grain was placed in each of three, 5 cm deep holes made 4 cm from the stem of each plant. Disease ratings were taken four weeks post-inoculation. Two DNA samples were taken per plant six weeks post-sowing and post-inoculation. DNA was extracted, quantified, and then combined with newly designed Ph coupling and repulsion qPCR marker assays to obtain genotype data. Disease ratings were compared to genotype data across two independent experiments and results indicated a strong correlation of genotype and resistant phenotype to Black Shank Race 0. When compared to the RAPD assay, the qPCR assay provides higher throughput in determining genotypes allowing for enhanced productivity in a plant breeding program.

Influence of maleic hydrazide on transcriptome landscape of chemically topped Burley tobacco

PATTANAIK S.; SINGH S.K.; RICHMOND M.; PEARCE B.; BAILEY A.; YUAN Ling
University of Kentucky, Lexington, KY, USA

Topping is a common agronomic practice in tobacco production by removing the top of each tobacco plants in an entire field. Topping stimulates proliferation of axillary shoots (a.k.a. suckers) that affects yield and leaf quality, thus requiring application of suckercides that inhibit sucker formation. Therefore, "chemical topping", by which the untopped plants are treated with a suckercide, is of interest as it eliminates the need for topping. Maleic hydrazide (MH) is a systemic suckercide widely used in tobacco production. The influence of MH on tobacco gene expression and molecular mechanism of MH action are not well studied. We aimed to study global changes in gene expression in apical (ApB) and axillary buds (AxB) of "chemically topped" burley tobacco using RNA-sequencing (RNA-seq) and to explore the molecular mechanism of MH action on sucker control. MH treatment significantly altered gene expression in ApB and AxB compared to the control. Gene ontology (GO) analysis revealed that the differentially expressed genes (DEGs) in ApB and AxB are enriched for signal transduction, meristem maintenance, cytokinin metabolism, cell wall synthesis, photosynthesis, and DNA metabolism. Furthermore, genes related to biosynthesis and signaling of a number of phytohormones, including cytokinin, jasmonic acid, ethylene, abscisic acid, and gibberellic acid, were affected by MH-treatment in ApB and AxB. We propose that altered expression of genes related to phytohormone metabolism and signaling, meristem development, and cell division, as well as DNA repair and recombination, likely leads to the inhibition of apical and axillary shoot growth. Collectively, RNA-seq analysis provides insights into transcriptome landscape of chemically topped tobacco and possible molecular mechanism of MH action on sucker growth in tobacco.

Effect of light irradiation on metabolite dynamics in tobacco leaves during curing

KOIKE T.; NAGAI A.
Japan Tobacco Inc, Yokohama, Japan

The components in leaf tobacco change significantly during curing. External environmental conditions during curing, such as temperature and humidity, significantly affect these changes. However, the effect of light irradiation on changes in leaf tobacco components during curing remains unclear. This investigation aims to identify the tobacco leaf components changed by light irradiation during curing. Tobacco leaves were cured under light irradiation or dark conditions at a constant temperature and humidity. Changes in the tobacco leaf components under each curing condition were measured. The measurement results are reported in this presentation. Flue-cured Virginia tobacco, cultivated in Japan, was used in this study. The tobacco leaves were cured under light irradiation or dark conditions, both at 38 °C and 60% relative humidity. Volatile components and nonvolatile components in the tobacco leaves were measured after curing for 6, 18, 30, 48, and 72 h at each condition using chemical analysis. All peaks of volatile component groups and nonvolatile component groups obtained by chemical analysis were subjected to principal component analysis. From principal component analysis, the component groups contributing to the first and second principal components showed no changes due to the presence or absence of light irradiation. However, a change was observed in the component group contributing to the third principal component. These results suggested that light irradiation of tobacco leaves during curing could induce changes in some metabolites and chemical components.

Key factors in enzyme deactivation during preservation under low-moisture conditions

KOBAYASHI K.(1); KURODA Y.(2)
(1) Japan Tobacco Inc., Yokohama, Japan; (2) Tokyo University of Agriculture and Technology, Tokyo, Japan

In general, the deterioration of foods and luxury goods must be suppressed until they reach customers. Drying is one such method used for this purpose. In the tobacco industry, the suppression of successive changes in the components of tobacco leaves that occur after curing is also important. Various enzymes remain in cured tobacco leaves, with their catalytic reactions known to cause changes in various components. Enzymes in cured tobacco leaves deactivate over time, but the main factors and mechanisms of deactivation are not clear. To improve the quality stability of tobacco leaves, a comprehensive understanding is required. Therefore, this study aims to elucidate the key factors affecting enzyme inactivation during cured tobacco leaf storage. In this study, we focused on temperature and water activity as factors related to enzyme inactivation. Burley tobacco leaves were used as samples and storage tests were performed immediately after curing. Samples were stored at 20, 40, and 60°C for up to 12 weeks with water activities adjusted to 0.1–0.6. After storage, the polyphenol oxidase, invertase, and glucosidase activities of samples were measured, with samples preserved at higher temperatures showing significant rapid inactivation. Lower water activities in the samples resulted in decreased enzyme deactivation rates. In this presentation, we report detailed results of changes in enzyme activity during storage and the effects of different temperatures and water activities on enzyme deactivation.

AUTHOR INDEX

AUTHOR INDEX

AUTHOR INDEX